GUIDE TO HOTEL SECURITY

GUIDE TO HOTEL SECURITY

Denis Hughes

Gower

Published by
Gower Publishing Company Limited
Aldershot, Hants, England

Printed and bound in Great Britain
by Billings & Sons Limited, Worcester.

British Library Cataloguing in Publication Data

Hughes, Denis
  Guide to hotel security.
  1. Hotels, taverns, etc. — Security measures
  I. Title
647'.94'0289   TX911.3.S4

ISBN 0-566-02428-4

Contents

Preface   vii

1 **Introduction**   1

2 **Organising a security department**   5
   Security policy – The chief security officer – Other
   security staff

3 **Hotel building security**   17
   Hotel design – Exterior entrances – Interior security

4 **Handling cash**   25
   Guests' money – Wages – Public telephone boxes –
   Cash taken at night – Control of procedures

5 **Guests and undesirables**   31
   Identification and treatment of undesirables –
   Control of guest and staff entry

6 **Security of rooms**   43
   Prevention of room thefts – Investigation of losses
   from rooms

7 **Security of property**   55
   Limiting legal liability – Facilities for safe deposit –
   Lost and found property

8 **Control of food and drink**   65
   The need for security systems – Bars – Food and
   beverages

9 **The housekeeping department**   79
   Room cleaning – Other housekeeping staff and
   services

10 **Other hotel services**   89
    Car parks – Pageboys or bellmen – Doormen – Hotel
    switchboard – Television sets
11 **Fire**   97
    The dangers of fire – Fire precautions – Action in the
    event of fire
12 **Terrorism and disasters**   113
    Bombs and other terrorist attacks – Disasters
13 **Training in security**   127
    The chief security officer – Other hotel staff

Appendix: The law relating to hotels in the UK   135
    Fire – Obligations of the hotel proprietor –
    Discrimination – Loss of or damage to guests'
    property – Safety of the guest – Registration – Trade
    descriptions – Right of lien – Criminal law – Obtaining
    services by deception – Arrest

Index   149

Preface

Hotel management is a skilled occupation and the day-to-day running of a successful hotel is both time-consuming and demanding of physical effort. The hallmark of the successful hotel is the 'No vacancy' sign, but success can soon turn sour if the hotel is beset by problems.

Many of these problems come under the heading of security, and security is a very important part of hotel management. Unfortunately it is an aspect that is often overlooked in management training or else only touched upon. This means that when these problems do arise the manager has difficulty in dealing with them, particularly if he has no professional security advice to call upon.

The object of this book is to fill the gap in management training and also to act as a handy reference for the hotel manager or the hotel security officer. He will find that many security problems can be solved by the use of common sense, but he must be able to recognise them, and having recognised them be able to use his knowledge to good effect. This book will give him the necessary background, increase his security awareness and equip him to handle the most common security-related issues. It will also provide assistance to those engaged in training managers.

Denis C. Hughes

1 Introduction

The hotel is probably vulnerable to more security problems than any other building. Running a hotel, whether large or small, involves allowing on to one's premises a considerable number of people, the majority of whom are unknown to the hotel owner, and who may pose a threat to the security of the property. To this must be added the risk of fire, with all the loss of life and damage to property that may result. It is only ten years or so since a fire in a hotel in Korea caused the death of over 150 people, and two years since a fire in Los Angeles caused eighty deaths.

It follows that any person responsible for the efficient running of a hotel must be aware of the security aspects of his task. Much of the liability is imposed on him by various pieces of legislation, not the least of which are fire regulations. The hotel owner also has legal responsibilities to those in his employ as well as the accepted responsibilities for the wellbeing and safety of his guests. The hotel manager often fails to realise the effect that a security problem has upon the running of his business. Too often a series of room thefts etc. is shrugged off as 'one of those things', and the cumulative effect of such crimes on the occupancy of the hotel is ignored. A full hotel usually denotes satisfied clients and can only be obtained by full attention to all the needs of the clients and by prompt action when complaints are received. The successful hotel also relies upon recommendation from satisfied clients, which will not be forthcoming if the client has been the victim of a room theft or other lapse in security. He himself will probably stay elsewhere in future and he will also tell all his friends and colleagues of his misfortune, no doubt blaming, sometimes without justification, the lack of security at the hotel.

Hotels have been with us for a very long time, and indeed were known to the Romans. In the early days in the United Kingdom this type of catering was carried out by the inn or tavern. By 1883 such premises were controlled by the Innkeepers Liability Act and defined as 'Any hotel, tavern, public house or other place of refreshment, the keeper of which is responsible for the goods and property'. By the nineteenth century the law placed responsibility for the goods and property of the guest, and therefore responsibility for security, firmly upon the innkeeper. In addition it was clearly understood that unless the house was full a traveller could not be turned away, provided that he was sober and in a proper state to be received.

In 1956 the Hotel Proprietors Act was passed, and it is this Act which governs hotels in the UK today. The Act will be dealt with in fuller detail in the Appendix but it is sufficient to say at this stage that this Act recognised that the hotel rather than the inn had become the focus of occupation, although conversely section 1 of the Act did say that a hotel within the meaning of the Act was deemed to be an inn.

The last two decades have seen an immense increase in the number and type of hotels available, and many of the so-called motel type have been built to cater for motorists, particularly those using the motorway networks. However, whether the building uses the name 'hotel' or 'motel' the problems remain the same, although it is probably true to say that catering for the motor car at hotels has led to additional security problems.

What are the security problems of the modern hotel?

They probably fall into four main groups: fire, theft, fraud and misuse. Fire needs little introduction. Its hazards are well known, or should be, and any hotel manager who has not studied this problem, paid attention to the fire legislation and ensured that his staff are properly trained, is in for a few nasty shocks. The heading of theft covers all crime that results in theft, whether a straightforward theft from the lobby, from a room or a public restaurant etc., or a forcible entry to a room or a robbery in the true sense of the word. Fraud covers the offences known in the hotel trade as 'walk outs', i.e. guests who leave hotels, restaurants and bars without paying, stolen credit cards etc. used to obtain credit, the fraudulent obtaining of guests' property, and similar offences. The last group heading, misuse, covers misuse of the building itself by persons not desired by the management, such as prostitutes. This problem will be dealt with in full in chapter 5.

It will thus be seen that the security problems that may arise in a hotel are many and various. A hotel security operation has been likened to the job of a village constable, and this is not a bad analogy. A medium-sized hotel has a population of perhaps over 1,000 people, a number of small individual dwellings, a public house type operation (the bars), the local café (restaurant) and perhaps a cinema or theatre. It is also probably fair to say that of the two jobs the village policeman has the easier one: he is dealing with a static population who probably never cause the number of problems the average hotel manager has to face, and he will certainly not be dealing with a selection of visitors from nearly every country in the world, many of whom will not speak his language, particularly when suspected of a crime!

No hotel is immune from security problems, and the small hotel in a small country town is just as liable to attack as its large city counterpart, although the proprietor may be less aware of the risks, and may not even know he has been attacked. Most of the large hotel chains are aware of the problem and have appointed a security adviser, and, at their larger hotels, a security staff. However, it is often felt that, just because a hotel is 'out in the sticks', problems will not occur. This is one of the clearest examples of 'hiding your head in the sand'.

Crime, even in the hotel world, knows no boundaries, and fire is no respecter of property or persons: one can just as easily burn to death in a small country hotel as in a large one in a city. The problems are there, they have to be dealt with, and, given proper back-up and training, they can be dealt with successfully.

2 Organising a security department

Whether a security department exists in a hotel will depend to a large extent upon the size of the hotel. It is obvious that a small hotel with perhaps twenty bedrooms will not need to go to the expense of a chief security officer and patrolling security officers. However, in very many hotels such a department is essential; even many of these, sadly, do not have one.

For the very small hotel where a whole department is not necessary it is still essential that one person is nominated to be responsible for security. At the extreme end of this type only the proprietor himself can take this responsibility, but as a hotel increases in size so the management's responsibilities must be split, and at this stage one person must be nominated responsible for security. If there is just one manager answering to the proprietor, he will be that person. If the hotel is large enough to warrant more than one manager, the manager with responsibility for reception, cashiers etc., usually called the front of house manager, is the most suitable person. In the very large hotel there should be a chief security officer who reports direct to the general manager. Responsibility for security is his, but in his absence the duty manager will deal with security matters.

Whatever the size of the operation the person responsible for security must be aware of his function, which must be made clear also to the rest of the staff. It must not be a responsibility assumed by just anyone who happens to be about at the time.

Security policy

The responsibility having been decided and placed very firmly upon one person, the hotel management's next duty should be to draw up a policy on security matters. This policy should cover all important aspects of security including fire and safety and should give clearly defined lines for the nominated person to follow.

One very important aspect of any security policy is the decision making procedure. There is no worse situation than when an incident happens and no one can decide what action should be taken, or indeed who should make the decision. For example, if a bomb threat call is received, who is going to decide whether or not to evacuate the hotel? If there is a fire, who is to decide at what stage evacuation should be considered? If a theft is reported, who informs the police, and are the police to be informed of all thefts, however small? If a member of staff is involved in a theft of hotel property, what is the policy as regards prosecution? These are only a few of the vital questions that need to be answered in the procedure laid down in hotel security policy.

Such a policy should, therefore, cover the following:

Fire precautions: fire warning system
fire fighting equipment servicing
escape procedures
fire fighting
liaison with fire prevention officer

Bomb threats: system for dealing with calls
evacuation responsibility
evacuation procedure
search teams

Burglary/theft	from rooms – reporting procedure in staff areas – procedure from other areas – reporting procedure hotel property – policy thefts by staff – prosecution policy
Room keys:	loss – reporting procedure control of master key systems replacement of locks policy control of key issue
Guests' property:	policy on security extra precautions for valuables lost and found property-procedure
Public areas:	control access by undesirables – procedure for dealing with drunks, prostitutes, etc.
Fraud:	procedure to be followed for 'walk outs' procedure for checking credit cards, etc. procedure for changing currency
Luggage rooms:	means of dealing with guests' luggage
Safety:	management responsibility check on hazardous areas

The above gives an outline of the various aspects that a security policy should cover. The list is not, of course, exhaustive, but should provide a good basis for designing such a policy, and will give good support and guidance to the nominated person.

The chief security officer

Earlier in the chapter it was stated that large hotels will

have a chief security officer, probably assisted by patrolling security officers. Where this is so the guidelines for the department should be clearly set out and a job description drawn up for the chief security officer.

The appointment

The appointment of the chief security officer is an important one and should not be undertaken lightly. Great care should be taken in selecting the right person. The first steps in this will normally be taken by the personnel department, who will issue an advertisement. What sort of person are they looking for? This question is best answered by the job description allied with the age scale and type of experience required.

The job description will vary to a certain extent depending upon the size of the hotel and the size of the security staff, if any, but it should be based upon the security policy set out earlier in the chapter. It thus follows that the person selected should be able to deal with crime, fire and safety and should have experience in at least one but preferably all of those fields. He will also be expected to take an active part in the security training of all hotel personnel, which should form part of the induction training, and take part in fire training. He should be a mature person with common sense and above all honesty and integrity. He should be capable of putting a forcible view of security to senior management, and be able to liaise with local police and fire departments. He should have a knowledge of the law as it affects hotels, and be aware of the many and varied problems facing the hotel security officer.

A police, fire service or armed services background can be an advantage, but is not essential, although the majority of chief security officers come from this group. Membership of a professional security organisation is an asset. The most important professional organisations in the UK are:

Guild of Hotel Chief Security Officers
Association of Hotel Chief Security Officers
Institute of Industrial Security
Institute of Professional Investigators
International Professional Security Association

In the USA the main professional body is the American Society of Industrial Security (ASIS); in addition, each state has its own security organisation.

Membership of any of these organisations shows that the applicant has taken the trouble to learn more about his profession, and in the case of the first two will also show that he or she has experience in hotel security, as it is not possible to obtain membership without the relevant employment. The two institutes are more academic bodies and set examinations, so membership or fellowship of these will indicate willingness to study and progress in the industry.

The personnel department will endeavour to obtain applicants of this quality and will advertise to that end. National newspapers can be used, but professional magazines (e.g. in the UK *The Police Review*, *Security Gazette* and *Security and Protection* and in the USA *Security World*) can be useful vehicles for such an advertisement. A short list of suitable applicants should be prepared for final selection by the general manager. It is most important that the final selection should be made at this level, as this is the level to which the chief security officer will report. He should make impact on the general manager, so it is essential that the two should understand each other. Finally it should be emphasised that there are many female security officers in the industry, who should also be given extremely careful consideration; no pre-conceived ideas that it is 'a job for a man' should be allowed.

It is not a function of this book to tell general managers how to conduct their interviews, but it is important that they should ascertain from the candidate how he will approach the job and what he knows about it, and above all ensure that he is not too rigid in his attitudes, but prepared to stand back and look at the job from a wider viewpoint. The manager should beware of the applicant who claims to know all there is to know about hotel security, and to have an answer for all the problems. Such a man or woman does not exist. Hotel security is like all security and police work: there is continual change, and work and study are needed to keep abreast of current trends.

The position

Once the chief security officer has been appointed it will be his task to ensure that the security of the hotel is as tight as possible, and to set up, if it does not already exist, a satisfactory security department.

The chief security officer will report direct to the general manager and should be classed as a department head, on a level with the front of house manager, food and beverages manager, head housekeeper, etc. The CSO should attend any department meetings, as very many of the subjects discussed will impinge on security. Usually the CSO will only be responsible for security officers, but to this responsibility may be added other members of hotel staff such as the timekeepers on the staff entrance, firemen if any, and sometimes such staff as goods inward.

The CSO will, in line with other heads, be responsible to the general manager for the efficient running of his department but another important part of his job will be liaison with the other department heads. Each department has its own importance within hotel management, but all must work together to ensure the smooth running of the hotel. A department of particular importance to the CSO should be housekeeping, which the wise CSO will regard as his eyes and ears. Of all the staff the housekeepers, with their force of chambermaids, etc., are the ones always likely to be on the floors and to see suspicious persons and report suspicious incidents. A good liaison between security and housekeeping can only be of benefit to the hotel.

Other security staff

Patrols

The first problem for the CSO on appointment is how to carry out his security function in its narrowest sense, that is, the security patrols. How these are carried out, their frequency and the hours covered will depend upon the number of security staff he has available, and the financial constraints of his security budget. Ideally a hotel should be patrolled twenty-four hours a day, but the cost of this could be prohibitive if the problems do not justify it and a suitable compromise may be a cover from 7 am to midnight and minimum cover at weekends. The CSO can consider this minimum cover where he is responsible for the type of hotel used by businessmen from Monday to Thursday and which then runs down for the weekend, usually endeavouring to fill up by the use of weekend budget breaks, etc. The security cover should be geared to cover the maximum occupancy times of the hotel.

One of the options open to the CSO when providing the security cover considered necessary is to examine the feasibility of using contract security for the patrol work, rather than employing his own staff. The main advantage of this is probably cost, but once cost features too highly in any security operation, efficiency takes second place. For example, it would be possible to obtain security cover from a contract company at a very low hourly rate, but what sort of staff would be patrolling the floors? Certainly contract security is a serious option, but under no circumstances should the contract be awarded to the lowest tender. There are advantages to both types of security and these could be summed up as:

Hotel's own staff: Own employees under complete control of CSO and hotel management. They can be placed on any duties considered relevant to security.
These staff become very familiar with the hotel, its security policy and its general operations.
The CSO retains complete control over who is employed on these duties and can fix his own standards as to the ability, the educational standard, background and qualifications of these staff.

Contract: An assurance and indeed a contractual liability that security staff will cover the duties required, obviating any need for the hotel to concern itself about sickness, holidays or absenteeism.
Staff that are unsuitable for the duty can be changed immediately by the

company without problem to the
hotel.
Security officers should be provided
fully trained and immediately able
to carry out the duties required.

When the level of security has been decided, and the
hours to be covered fixed, then the duties of the
patrolling officers should be clearly defined. The normal
patrol duties would consist of the following:

To make regular rounds of all hotel corridors and
areas, including public rooms.
To observe any suspicious or abnormal activities on
the part of either staff or guests.
To be alert for undesirables such as drunks and
prostitutes and if necessary evict them from the
premises.
To assist the duty manager in inspecting the rooms of
doubtful clients and assisting with suspected 'walk
outs'.
To commence investigation of any offence reported
whilst on patrol, and to ensure that full details are
obtained for subsequent report to the police where
necessary.
To ensure that staff, particularly maids, conform to
rules concerning shutting of guest room doors whilst
work is in progress.
To look out for room keys left in doors.
To seize and deal with property found on patrol.
To maintain a close watch for any fire or safety
hazards.
To make a full report at the end of each patrol on any
incident that has occurred.
To provide escort where required for cashiers and to
assist in the emptying of telephone boxes etc.

Again this is not an exhaustive list but an outline of the

probable duties of a patrolling security officer. The key word is alertness, and a good officer will be alert to notice anything that is unusual or amiss.

The style of dress of the patrolling security officer must be a matter for the individual hotel. The author favours a smart blazer and grey flannels, which clearly indicates that the person is someone special without being too obtrusive. Many hotels use officers in normal clothing, while others use officers in full uniform. Each practice has its advantages as well as disadvantages, and only experience will provide the answer. Patrolling security officers should, like all security or police officers, act as a deterrent, and it is difficult for them to do this if it is not apparent who they are: on the other hand, if they are not obvious they may stand more chance of catching the casual thief.

Cash collection

The one area where the use of contract security is certainly desirable is the collection of cash from the hotel for payment into the bank, and the collection of wage money from the bank. No hotel management should put its own staff at risk for this duty and a CSO should resist any attempt to involve his staff. This duty should be handed over to a competent company with the proper vehicles, containers and uniformed men equipped to carry out the task. The cost of this service will not be excessive, and will ensure that the cash is carried at far less risk to the hotel and its staff.

Attention to the points made in this chapter should facilitate the formation of a good, efficient and competent security department. The role of the individual appointed to be responsible for security in the hotel is extremely important, and the utmost care should be taken in making this appointment.

3   Hotel building security

A hotel building can be one of many types but the main classes are: the modern, purpose-built hotel, the modern purpose-built motel, the older but now modernised hotel, and the large building converted into hotel accommodation. The security problems are, in essence, the same for all of them, but there will be a greater or lesser number of problems as the design of the building dictates.

Hotel design

The most secure type of hotel building should, in theory, be the modern one, drawing on the architect's long experience in designing hotels and a familiarity with the security problems encountered in older buildings. Unfortunately this is far from the truth. In the construction of many modern hotels priority has, quite rightly, been given to ease of access and the convenience of the prospective client. The security of the guest and his property, and the security of the property of the hotel itself have, however, been disregarded. The security problems faced by hotels have already been mentioned, including the offence known in the hotel trade as 'walk outs'. An effective guard against these offences, which can cost a hotel a considerable sum, is to have properly trained reception and hall porter staff who will be on their guard against people slipping out without paying and will keep a close eye on lifts, stairs, doors etc., to that end. Yet the author has visited many modern hotels where reception is on the first floor and there is no check at all on the ground floor. Another hotel visited gave absolute access to parked cars from each floor, a helpful facility, but for whom? Yet another gave access by staircase and lift direct to the underground car park – highly desirable, except that at the time large quantities of sheets and linen had been stolen, and the TV sets were starting to go!

Of course due thought must be given to ease of access by guests, but by the same token thought must also be given in the design stage to the problems that this may cause. And is a reception service on the first floor really of benefit to the guest?

It follows, therefore, that great care must be taken in the design of a modern hotel, and that the advice of a security expert should be sought at the very earliest stage. Any security measures, particularly electronic ones, are much more cost effective if installed at the time of construction and not as an afterthought. If a car park is to be underneath the hotel with direct access inside, then perhaps two sets of lifts are necessary, one to the ground floor and then another to the rooms, so that direct access to cars with an armful of hotel property is not possible. Or the use of closed circuit television (CCTV) can be considered, with the cameras placed in a very overt position so that the security is obvious and acts as a deterrent.

Exterior entrances

In reality, however, hotel building security tends to be discussed only after the building has been erected, when a number of different features should be considered. A starting point should always be the exterior entrances: a hotel is unusual in that the exterior entrances must be available for access at all times. Unless a hotel is very small it is never securely locked up, and therefore effective security at entrances is not a straightforward matter. However, there will be many entrances that can and should be secured at all times, including fire exit doors, staff entrances, delivery points and other odd doors not used by the public for access to the hotel. Care must be taken when dealing with fire escape doors that any measures recommended do not cut across the requirements of the fire regulations. This problem can normally be overcome by securing these doors with the type of fire panic bolt known as a Redlam bolt, a panic bar kept in a locked position by a bolt and padlock. The padlock can be unlocked by a key, but in an emergency it is easily released by the simple expedient of breaking a glass tube forming part of the bolt. These bolts give excellent security whilst at the same time complying in every way with fire regulations. An additional measure that can be considered is the fitting of a simple alarm wired to, perhaps, the timekeepers' office, so that any interference with the door is immediately notified. Such a measure is important as it has not been unknown for these doors to be used by staff wishing to remove hotel property. One team of two even had their own supply of glass tubes! Additionally, such doors should be included in a regular security patrol.

Public entrances

The public entrances to the hotel are always extremely difficult to control. Ideally the doors should always be in the view of hotel staff, particularly the hall porters and reception, who must be security conscious. Very often access to the hotel can also be obtained through restaurants and bars and it is imperative that this access is controlled or even stopped late at night. Restaurant and bar doors leading to the street should be secured when normal hours are finished. It should be a golden rule that after midnight the entrance doors available to the public are kept to a minimum.

Many hotels now have CCTV permanently focussed on the main entrance with a video recording and playback facility which may often help to identify a thief. CCTV is a great aid to security but should never completely replace the trained observant member of staff, who provides the most effective security there is.

The staff entrance

The other entrance which is open twenty-four hours a day and therefore needs to be controlled for the whole of that time is the staff entrance, which should be controlled by timekeepers with some security training. A word of warning is necessary here. Too often the timekeeper's job is given to the elderly or unfit, who are not suitable for this task. It is a job that could be seriously considered for contract security, as above all the timekeepers must be efficient and retain an independence. This timekeeper's entrance is another area that must be subject to frequent patrols by security officers.

Interior security

When the exterior entrances of the hotel are satisfactorily controlled, the attention of the staff can be turned to the interior of the hotel. The various functions of a hotel will be dealt with in more detail elsewhere in this book, but here we should look purely at physical security throughout the building.

The main security concern of any hotel must be for the guests, the rooms they are occupying and the property therein. From the physical security viewpoint, the doors to guests' rooms must be as secure as possible. This particular problem is covered very thoroughly in chapter 6. Security of rooms is paramount and must not overlook two further aspects, the access to the rooms through corridors etc., and any windows in the room itself.

Windows

Unfortunately windows are often ignored as a security risk, particularly when they are not on the ground or first floor and so are considered out of reach. Nothing could be farther from the truth. A friend of the author, staying at a hotel in the USA, occupied a room on the sixth floor. His room overlooked the street with a sheer drop. He retired to his room at about 10 pm and opened the window to let in some air. He was lying on his bed sipping a whisky when suddenly he noticed his curtains moving. To his horror a large hand then appeared grasping a wicked-looking knife, and a man climbed into the room. He approached the bed in what could only be described as a threatening manner and the author's friend left the room in such a frenzy that he found out later he had torn the chain from the door. Fortunately he ran straight into a patrolling security officer. When they returned to the room the man had gone. The victim was told by police that he was very lucky, as without doubt if he had stayed and tried to defend his room the man would have used the knife. When a query was raised as to how the man had gained access up a sheer wall to the sixth floor, the reply was that this was not uncommon. The point of this story is that windows of hotel rooms are very important even when the room is situated well away from the ground. Many hotel thieves are good climbers!

The security of windows should, therefore be examined in some detail, and a way should be found to arrange circulation of air without making the window a means of access to the room. Much will depend upon the type of window and its vulnerability; for example, a window overlooking a balcony will be more at risk than one facing a sheer drop. Fortunately, in the UK at any rate, most thefts take place when the room attacked is unoccupied, so it is extremely unlikely that the type of attack described above will take place in the UK. In that country the securing of hotel windows will generally be a securing when the room is unoccupied, and this is probably best achieved by the use of simple and cheap window locks.

Corridors

The corridors leading to rooms also merit attention and, where there is a security patrol, they should be visited at frequent intervals. Even where there are patrols and certainly where there are not, the use of CCTV should be seriously considered. The actual presence of the cameras themselves is a deterrent, and enables management or security to make frequent checks of vulnerable areas in a very short space of time. This system can also be used as a means of supervision, particularly of room cleaning staff who do not always follow strict instructions to secure doors whilst rooms are being cleaned. The initial cost of installing the cameras will be high but the cost can be justified, not least in that guests can see that the hotel is taking every possible step to ensure proper supervision from a security angle. Finally, such a system can also be a back up to a fire patrol, especially in those areas not effectively covered by smoke or heat detectors – remembering that even a detector can be faulty.

4 Handling cash

Quite a considerable amount of hotel crime does not affect guests' rooms at all, but other parts of the hotel, such as cashiers' and luggage rooms. Most hotels will have three separate operations dealing with cash: firstly the cash taken from guests when checking out, etc., secondly the cash handled in connection with staff wages, petty cash etc., and thirdly the cash from telephone boxes. Usually these deposits of cash are housed in separate locations, which need attention from the physical security viewpoint.

Guests' money

The cashier's room dealing with guests' money is usually located near reception and in many ways is very vulnerable to attack, though strangely enough it does not often find itself at the receiving end of crime. Two basic security measures should be taken here, namely proper control and the supplying of suitable receptacles for the cash such as safes etc.

Control should be exercised by the chief cashier, if there is one, or, otherwise, by the duty manager, with spot checks by the CSO and the general manager. Every cashier should have his or her own float which is checked before and after their tour of duty. The cash point should be supplied with a slam lock till to ensure protection against a till snatch. When the contents of the till are removed at the end of the tour of duty they should be immediately placed in a container and locked in a suitable level of safe, such as a Treasury type, which is adequately covered by insurance. The key should be in the sole and firm control of one person with a duplicate held in a safe in the security office, or the general manager's office if there is no security. Where the hotel changes quite large sums of foreign currency a number of spot checks should be made on cashiers and the amount of currency held, in order to safeguard against a cashier setting his own rates of exchange, which he can then use to his advantage with the result that both the client and the hotel lose.

The other areas handling guests' cash are the bars, restaurant and lounge service. For reasons to be outlined later, on no account should floor service staff handle cash. It is not possible or indeed feasible to have such a rule in the bars and restaurants, so cash handled in those areas must be safeguarded. Firstly, where tills are used, they should be of the latest type and again should be checked before and after each set of working hours. Managements and security staff should make frequent checks in these areas and observe the working of the tills. The most effective supervision of these operations is by a proper control system of documentation, and this will be covered in chapter 8.

Wages

We shall now turn our attention to the cash held in the cashier's department dealing with staff wages etc. Ideally the money will be delivered by a cash carrying company, but where will it go? How secure is the pay office? Where is it situated? These and many other questions will need to be answered before one can be satisfied that the cash is being securely handled.

The cash for wages should preferably be made up into pay packs by the delivering company so that they can deliver on the day of the pay out, thus reducing the time that the cash has to be housed on the premises. The cashier's office should be situated away from the ground floor and from any staircase or lift. The office itself should be of secure construction with a solid door, steel lined, fitted with two mortise locks of at least five levers' and an observation hole. The door should be linked to an alarm system. The office should also have a good safe capable of holding the amount of money likely to be left overnight. In addition to the alarm on the door the room should be fitted with panic alarm buttons at strategic points to cover a sudden and violent attack. These are particularly important at any pay out point. The pay out points themselves should not just be ordinary sliding windows or hatches but properly constructed and secured pay out points, with bandit-proof glass and a speak through facility. It should be a golden rule that no person at all, of whatever position in the hotel, should be allowed into the office after the cash has been received and before it is paid out. It should not have to be stressed that the staff employed in these offices should be honest, trustworthy, reliable and properly vetted by the personnel department before employment.

Public telephone boxes

The last area of cash handling that needs to be examined is the emptying of public telephone boxes. In a large hotel a considerable sum of money may be involved, and unless it is properly dealt with this money may be put to uses other than those intended.

These boxes should be emptied on a regular basis and always by at least two staff from different departments, one of whom should be a security officer if one is available. The money should be emptied straight into a suitable container, and then taken back at the end of the collection to an office and there counted out, checked and then counterchecked by a third person. This procedure sounds strict, but this area is often overlooked, in which case someone will soon find ways of taking the money for their own use.

Cash taken at night

Problems can arise where the large hotel operates a cashier service during the night, usually during times when a minimum staff are on duty. At these times additional measures should be taken to protect the cash, and in addition to the slam lock till already suggested, a floor safe should be considered, which could then be used to safeguard money from the till once it had gone above an agreed maximum. Such a procedure would mean that, in the event of an attack, the amount obtained would be kept to a minimum and the risk to personnel reduced.

Control of procedures

Management or security should visit cashier's offices to ensure that the strict security rules are carried out. The author knows of one cashier's office which lost money out of the safe, and enquiries showed that the keys were left in the safe. When the hotel was visited as part of a security survey a few days after this event, the safe was again found to have the keys left in the door and money inside it! Remember that even the strictest rules rely upon the staff carrying them out.

It is fair to say that all cash handling jobs involve temptations for those responsible, and it is far better to ensure that only strict procedures are followed for such operations so that the persons employed do not succumb to these temptations, leading eventually to adverse publicity and the lowering of the hotel's reputation.

5   Guests and undesirables

A hotel can be classed as a building which gives almost unlimited access to members of the public while retaining the right to refuse admission to people it considers undesirable for one reason or another.

Persons entering hotels usually fall into one of the following categories:

a guest
a visitor to a hotel guest
an invited guest of the management
an employee
a visitor invited by an employee
an outside contractor
an undesirable – such as trespasser, prostitute etc.

The actual rights of each person will differ and the obligations owed by the hotel management will vary in respect of each category. At first sight a person walking into a hotel entrance cannot be categorised unless, of course, he is an obvious drunk, so selection will be made later. However, hotel staff who are experienced in such matters may make a preliminary mental selection as the person enters, and management are quite at liberty to refuse entry to any person providing that they have adequate grounds for so doing.

Identification and treatment of undesirables

As this book is dealing with security problems, it would

be as well to commence this chapter by considering the persons who are undesirable; the easiest of these to establish is undoubtedly the drunk.

Drunks

A drunk is without argument an undesired person who can be refused entry and certainly should be if desiring entry into a bar area. Also, it is a criminal offence to be drunk and incapable or drunk and disorderly in a public place. A problem arises when a bona fide guest of the hotel who has booked in quite properly goes out and then arrives back intoxicated. In this case it is better not to refuse admission but to establish his bona fides, and if feasible escort him to his room as quickly and with as little fuss as possible. However, this decision must rest with the duty manager, who must have in mind the wellbeing of his other guests.

A partly intoxicated person whose behaviour is somewhat of a nuisance, but not unlawful, may arrive at the hotel claiming to be a friend of a hotel guest and demanding admission. He should not, of course, be admitted, but the hotel could arrange to inform the person he claims to know whilst at the same time pointing out that in the management's view the visitor is in no state to be admitted to the hotel. This case calls for much tact and discretion.

Probably the easiest drunk to handle from the management side is the intoxicated member of staff who tries to come to work in that state. This will normally be a breach of his contract of employment, and this person should always be refused entry. In the same way, if a member of staff becomes intoxicated whilst working at the hotel he should be escorted off the premises.

Thieves

The most difficult undesirable to detect is the thief. He will not advertise his presence or his trade! The detection of such people relies solely on the keen observation and common sense of the staff. On floors the suspicious chambermaid is worth her weight in gold, and in the lobby the porters coupled with the link men or doormen will be the first line of defence. A particularly popular crime is stealing from guests in the foyer area, and anyone hanging about unnecessarily in this area should arouse suspicion. Strictly speaking their reason for entering the building means that such people are trespassers, and can therefore be invited to leave if detected, but this course may not be the best one in the circumstances. Hotel staff should be trained in this respect, stressing that people are not always what they seem to be, for example, that woman dressed as a chambermaid might not be a chambermaid. This is of great importance when dealing with such outside contractors as window cleaners. One could almost guarantee that if one were to enter a hotel in overalls with a leather and a bucket and a confident air, one could walk round the hotel without let or hindrance. It should always be brought to the attention of staff that thieves often enter in this guise, and they should be taught to question anyone who appears to be a member of staff or a contractor, but whom they do not know.

'Walk outs'

Equally undesirable is the potential 'walk out', but he is without doubt the most difficult to detect. Not all hotels are so fortunate as to boast a manager who can 'smell' out a potential 'walk out' before he books in. There is a good early warning system in large cities from hotel to hotel, as such people do usually make a habit of the offence, but in country areas this early warning system may not exist. To leave without paying is, of course, a criminal offence usually reported to police, so persistent offenders will be circulated by the police force concerned. Any such descriptions received at hotels should be thoroughly studied by reception staff and where possible posted prominently near the staff in the reception area. Many potential offenders will not want to give full details of themselves, and certainly not the correct details, so reception staff should train themselves always to ensure that whatever medium they use for registration is filled in completely and properly. Far too often following a 'walk out' one has seen an incomplete registration form with so little detail on it as to give no assistance whatsoever.

There are one or two pointers that reception staff can watch for: lack of proper luggage; a certain hesitation in providing details for the registration; lack of evidence of identification. As a deterrent measure reception staff should always insist on a deposit or at least a print from a credit card from a person they do not know or have had no previous dealings with at that hotel. This does not obviate the fact that the credit card might be stolen, but at least it shows that the reception staff are alert. If a person wishes to settle by cheque a cheque card should automatically be required. The production of the actual cheque card is essential; in one reported case the offender claimed he had left his credit card/cheque card at home, but had a note of the number in his wallet.

The main problem that the hotel staff will find with these people is their plausibility — the main weapon in the armoury of any fraudster. Many also have the ability to look and act confident, so perhaps we should also be aware of the over-confident person. It is not easy to walk this tightrope of sorting out the potential fraudster. The last thing the conscientious member of staff wants is for his hotel to be defrauded, but at the same time he does not wish to upset a potential good client by treating everyone as suspicious. The whole answer to this problem is good training for such staff, not just in their duties but on the problems that they may be called to face and how to deal with them. This will, no doubt, add to the initial training costs of staff, but it will pay dividends in the long term.

Prostitutes

In the same way as the fraudster, the prostitute does not carry a sign advertising her trade, nor does she normally show any visible characteristic to enable staff to distinguish her from other women frequenting the hotel.

The first and most important step for management in dealing with prostitutes is to set a policy for staff to follow. It is not, of course, an offence just to be a prostitute; offences occur when prostitution is carried out by soliciting in the street, or where another person lives on the earnings of a prostitute, or when premises are used as a brothel. So the mere fact that the prostitute is on hotel premises does not, in itself, create an offence and she should be allowed all facilities available to other members of the public unless she decides to try to carry on her trade in the hotel. At that point she would become an undesirable guest, which would enable the staff to ask her to leave. It should be remembered that parts of the hotel, namely the bars, are licensed premises, and it is an offence to allow licensed premises to be the habitual resort or place of meeting of reputed prostitutes whether the object of their resorting or meeting is or is not prostitution (section 175 Licensing

Act 1964). However, there is a defence built into this section which states that the use of the premises for the purpose of obtaining reasonable refreshment is not prohibited. Hotel management should be aware that it is not unknown for members of hotel staff to have arrangements with prostitutes to have them available for guests, and that if this arrangement is with a member of bar staff and the prostitutes congregate in the bar for that purpose, the above offence could be committed.

One of the most difficult problems for hotel staff in this respect is with the bona fide guest who returns to the hotel with such a lady. What should they do? The danger from the hotel viewpoint is that when the lady has finished with the original client she may then try to ply her trade in other parts of the hotel, thus causing problems and, of course, bringing the hotel into disrepute. A hotel policy should be laid down to cover this situation, and a great deal of tact and discretion must be used by staff. One of the many dangers that can arise from too strict a policy in this respect is that a man may return to the hotel with a perfectly respectable lady, possibly even a relative, and be confronted by staff. This causes great embarrassment all round, and may cause even more harm to the hotel.

The whole secret of dealing with all undesirables in a hotel lies in proper training, allied with keen observation and the local knowledge of the staff. Where prostitutes operate in a particular area, they will soon become known to the observant member of reception or hall porter staff, and this knowledge is really the hotel's best defence against the too frequent use of the premises for prostitution. In a large city the problem will be greater and will never be stopped, but the staff's prevailing desire should be to ensure, as far as possible, that the hotel is properly run, retains a good reputation, and does not become trading ground for prostitution.

The right to refuse entry

When the background of hotels was examined in chapter 1 it was shown that by the Hotel Proprietors Act a hotel means an establishment held out by the proprietor as offering food, drink, and if so required, sleeping accommodation without special contract, to any traveller presenting himself who appears willing and able to pay a reasonable sum for the services and facilities provided, and who is in a fit state to be received. It is important to remind the reader that the law sets out very neatly the requirement to admit people to the hotel and shows that no decision to refuse entry to a person should be made lightly. There is no right to select guests, although the management may refuse to supply sleeping accommodation if all the bedrooms are full.

Control of guest and staff entry

Having now discussed undesirable hotel users, we must turn our attention to the bona fide users of a hotel, which will be everyone who walks through the door, unless there is evidence to the contrary. In a large hotel reasons for wanting to use the hotel may range from the normal requirement of sleeping accommodation through the use of bars, restaurants, coffee shops, lounge facilities, cinemas, to private functions such as banquets, conferences seminars, meetings, wedding receptions and a variety of others. Often the persons coming to the hotel will use only one of the hotel facilities, and very few will avail themselves of all that is available.

The larger the hotel the more people will be demanding admission, and to control this large number of people is an almost impossible task. It is at the times when the hotel is at its busiest, and bursting at the seams, that the hotel thief will take his opportunity to operate and, at these times, any security at the hotel should be concentrated on patrolling the floors away from the public areas to detect the thief and other undesirables.

In addition to the persons set out above as being bona fide users of the hotel, there will be the others we discussed briefly at the beginning of the chapter, namely the employees, guests of the management, guests of employees, outside contractors, persons making deliveries etc., and we shall now examine these classes in more detail.

Hotel staff

Employees should be known to one another, but in the early stages of employment this is not necessarily so, and in a large hotel with perhaps in excess of 500 staff it does not follow; some members of staff will never meet. The control of staff is important. The first step is to make a firm rule that all staff enter and leave by the staff entrance. This rule should be rigorously enforced, and apply at all times when staff leave the hotel, even if they only go for a short period. There should be a book or other record at the staff entrance for management level, and all other staff should be covered by a clock card system. It is equally important to have this entrance manned by timekeepers who should have the authority, covered by contracts of employment, to search the bags etc., of staff leaving the premises. Together with the requirement for a clock card or a book entry should be a requirement for all bags, holdalls etc., to be left with timekeepers and collected when leaving the hotel. Staff theft can be quite common in a hotel, and every step should be taken to ensure that the opportunities to commit this are kept to a minimum. Staff should not be allowed to carry to their department empty holdalls etc., which they can carry out later in the day, probably full! Another aspect of staff control, which is often overlooked is the necessity to be aware of the number of hotel staff on the premises in the case of a fire or a bomb attack. An essential part of any efficient evacuation procedure is a complete knowledge of the staff present on the premises, which can only be achieved by the methods referred to.

One final point to be made in this connection is the simple one of the control of hotel staff catering. The normal practice is to supply meals free to staff on duty. With a large staff the cost is already great and can be considerably increased by staff off duty; it has even not been unknown for persons not on the staff to obtain meals in a staff canteen.

The main control, and it does need to be stressed, is the insistence that staff enter and leave by the one staff entrance and no other. Only by this method can the required level of control be obtained and supervised.

Guests of hotel staff

Guests of the management should always be directed by their potential host to report to the reception area. Such persons should not be directed to the part of the hotel where the member of management staff is waiting, but should be asked to wait until collected. Only in that way can it be certain that the person is who he purports to be. Under no circumstances should guests of the management be allowed unrestricted access.

Guests of other employees should be invited to make their way to the staff entrance, where they should be collected by their host, although it would be advisable not to encourage many visitors of this type. If such a person calls at the front of house reception, he should be redirected to the staff entrance and wait there until seen by the member of staff. If he enters the hotel, he should be recorded as being on the premises.

Outside services

Outside contractors are quite commonly employed in a hotel, usually as window cleaners, plumbers, electricians, locksmiths etc. All such contractors must be strictly controlled; in most cases they will be admitted access via the staff entrance and their presence recorded by the timekeepers. Certain contractors, such as window cleaners, should have their agreed times of access to the hotel clearly defined, normally during the morning hours. This acts as a good safeguard against the bogus person dressed as a contractor, as anyone of that type seen operating in the hotel out of the normal times is immediately suspicious. Areas of work for contractors should be agreed, and where necessary a permit to work should be given. The responsibilities of workplace legislation should always be borne in mind. Where there are security staff, they should always be informed of contractors on the premises and the areas in which they are working. Where there are no security staff, the duty manager and any hotel staff likely to be on duty in the area of work should be so informed. Contractors should never be left unsupervised for long periods and should not be allowed to wander in and out of the hotel at will.

Deliveries form the last class of user of the hotel to be covered and are relatively simple to deal with. The rule should be for all deliveries, and indeed collections, to be dealt with at one point, which should be away from the main entrance, unless of course the delivery is an article for a guest. Articles delivered for guests should similarly be dealt with in one area; it is preferable for these items to be handled by hall porters. A word of warning is necessary here to ensure that, if this system is adopted, the porters not only have some method of physically recording the arrival of such goods, but are also given facilities for storing the goods until collection by the guests. Under no circumstances should such goods be handed over to anyone other than the guests, except by written authorisation. The person making a delivery, whether to the hotel or to a guest, should not be allowed past the area to which the delivery is made. Collections from the hotel should be dealt with as above and it only remains to emphasise that any collection should be closely supervised, as only the goods authorised should leave!

The foregoing paragraphs cover the majority of the users of a hotel, but not absolutely all. There will always be the casual visitor, through inquisitiveness, the visitor making an enquiry and many others, but the main classes are those referred to and attention to all of these should ensure that the hotel runs in an efficient manner.

6 Security of rooms

The one aspect of hotel security most likely to affect the guest, and which therefore needs the closest attention, is the security of the guest's room. It is from this room that the guest is more likely to suffer loss, and no doubt the state of the room, or its lack of cleaning, is more likely to lead to a complaint than anything else. The guest, quite properly, will expect to be able to leave certain property in his room and find it there when he returns, and will be shocked if this is not the case. If he does suffer loss it can also be expected that the last person he will blame for that loss will be himself. He will blame the hotel, its lack of care and non-existent security, and be most unlikely ever to visit that hotel again, nor is he likely to recommend it to his friends.

This introduction should prepare the ground for what is probably one of the most important chapters in the book, and one aspect of hotel security to which even the smallest hotel should pay the closest attention. It is fair to add that with a little thought and by the use of proper security measures the room can be made secure against anything except the very determined attack. In the latter case the hotel room stands very much as the private house, in that a determined criminal will obtain entry whatever security measures are taken.

Prevention of room thefts

Doors

It is stating the obvious to say that the easiest way to enter any room is by means of the door. The first requirement of security of a room must therefore be to provide a suitable door, one which will delay entry to the unauthorised visitor for as long as possible. What should not be overlooked in this respect are the requirements of the fire legislation, which should always be borne in mind when the physical security of a door is in question. The construction of the door itself is the first consideration, not forgetting the door frame, as it is useless having a door of solid construction in a frame that will easily give way. The doors should be of a solid construction, not the hollow core doors which are sometimes used. The construction of the door must be decided before the selection of the door locking system is made.

Connecting doors between rooms should not be overlooked, though they often are. They are often found in older hotels, and in larger hotels where rooms can be let off as suites. These connecting doors should be treated in the same way as doors off the corridor, as often the adjoining rooms are occupied by different guests and it is imperative under these circumstances that all doors are secure.

Locks and keys

The next consideration is the type of lock to be used on the doors; there are many types and varieties. The cost of the locks should not be a major consideration, but remember that the most expensive lock will not necessarily be the best. In this field the services of an expert are essential and whenever locks are considered the services of a locksmith should be sought. There are many firms manufacturing locks for hotels who will also be happy to give advice, but it should be borne in mind that this advice will not always be completely independent and that the person who comes to advise may also be a salesman.

The main drawback in providing a secure locking system for a hotel is the fact that access to the room needs to be obtained at all times, both by the guest and by hotel staff. It would be possible to recommend locking systems that made entry to the room impossible except for the guest, but what would happen if the guest in the room were ill? How would the staff gain entry? So it is necessary for any system used to allow access to a number of people. The most effective mechanical lock is without doubt the mortise lock, which can be set into the door with the tongue of the lock fitting into a metal striking plate set into the locking post of the door frame. There are many varieties, operated with a variety of keys, but all of them work on the same principles. Locks to avoid are the simple night latch commonly called the Yale type lock. These can be opened by a piece of plastic or a credit card and offer little protection. However, Yale do manufacture other very suitable locks, and any reputable lock firm such as Yale, Chubb, Banham, Ingersoll, etc., will always be prepared to give advice.

Room keys. The proper security of the door with a good lock is only the first part of an efficient security procedure. The next step is the control of the issue of the room keys themselves, and this is where many of the problems can arise. The keys to guests' rooms are handed over at the reception desk or hall porters' office when the room is first assigned. The guest then has the option of keeping the key in his possession for the duration of his stay or returning it to the desk when he leaves the hotel and picking it up later when he returns. Most hotels will recommend their guests to take the latter course, but many guests do not as they do not like the key going out of their possession. Some hotels, in an endeavour to enforce return of the key will attach a large tag to it, making the whole thing cumbersome for a pocket or handbag. The guest soon overcomes this by removing the tag from the key.

The first step in proper key control is to ask any person requesting a key to identify himself. Most hotels issue a card or booklet with the room number and guest's name to each guest as they book in. No key should really be issued until the card has been produced, but sadly this policy is not followed and, particularly in the larger hotels, it is quite possible to walk up to the key issuing point, quote a room number and be given the key without question, something which should never happen as it is very bad security. Reception staff who have been questioned about the practice will say that they learn to recognise the guests, which, quite frankly, is rubbish and should never be accepted by management. The duty manager should always supervise this operation closely and should not hesitate to intervene if he sees any slackness creeping in with the issue of keys.

The second necessity with key security is to ensure that keys kept at the issuing desk are never left unattended or indeed accessible to anyone reaching over the desk. All keys must be kept well out of reach and keys handed in or returned by guests should be dealt with immediately, not left lying on the top of the desk, as is often the case.

Thirdly, every effort must be made to ensure that guests leave their keys when they finally leave the hotel. A considerable number of keys get lost – some guests like to keep them as souvenirs – so guests must be politely asked when they check out whether they have handed in their key. The hotels who place tags on their keys usually indicate on the tag that the key can be returned to them by placing it in the nearest mail box; a convenient service, but also a convenience for other persons who may get their hands on this key. Therefore, if this system is adopted, not too much information should be given on the tag, so that the person obtaining the key will not know the hotel and the room it opens. The hotel thief for one would love to get his hands on these keys.

Lost keys. There must also be a procedure for dealing with lost keys and their replacement, and a policy must be decided for the changing of locks after keys are lost. In a very large hotel it may not be practical to change a lock every time a key is lost, whereas in a smaller hotel it will be. The simplest way to change locks is to have a certain number of spare locks for each floor and to replace the lock each time a key is lost. The replaced lock can then be used for another room on the same floor. Such a system means that, if the wrong person has got hold of the key, he will not be able to open the room shown on the tag with it. Of course, an experienced room thief will probably try other doors as well, but this takes time and thieves do not want to hang about in hotel corridors trying keys.

The issue of spare keys to rooms should not be a haphazard affair, but should be under the control of one person, preferably the security officer if the hotel has one. Enquiries should first be made to ascertain what has happened to the key, who lost it and when, and then the issue of a new key should be recorded and signed for.

Master keys. Whatever the locking system, it will be necessary to have a system of master keys, and with the issue of these arises the greatest security problem, for a hotel master key is equivalent to gold dust for a hotel thief. This system must be very strictly controlled and again this is a task that should be undertaken by security. If there is no security, then it is so important a task as to require the personal attention of the manager. The first question to be posed with any master key system is 'How many levels of keys?' and the second 'Who will be issued with them?' The answer to the first question will depend upon the size of the hotel and the policy decided. The levels that can be obtained will range from a grand master to a master, to a submaster, to a floor master and sometimes a section master, although generally speaking this last is unnecessary. The first level of key, the grand master, should be in the possession of only perhaps three people at the most: the general manager, the chief security officer and the duty manager. These keys will not only unlock every door in the hotel but should also 'double lock' the room. 'Double locking' is sometimes necessary where the guest has requested it for additional security, or where it is thought that a guest is a potential 'walk out' and it is thought necessary to lock the door so that he cannot get back in. In these cases no other key will unlock the door and one of the grand master key holders will need to unlock it when necessary. The next level, that of master, will unlock any room in the hotel but will not have the facility to unlock a double locked door. The submaster will unlock any door in a certain part of the hotel only, and again will not unlock a double locked door. The last type of key issued, the floor master, will, as its name implies, only open a door on a particular floor.

The master, submaster and floor master keys should be issued only daily to persons requiring them. They should be issued purely against signature and signed back at the end of that person's tour of duty. A book should be kept for this purpose and any loss should be reported and investigated immediately. The staff who are most likely to use master keys are the housekeeping staff, particularly the chambermaids, who must be instructed to fix the keys to the chain or key strap which is normally issued to them. It should also be emphasised to these staff that, if they are approached by a guest who claims to have lost his or her key and asks to be let into the room, they should not let them in, but call for the attendance of the patrolling security officer, if available, or the nearest housekeeper.

Electronic systems. The modern trend to counteract many of these problems is to move to an electronic locking system. Instead of a key to a door the guest is issued with a card which is inserted in a slot in the door in the normal position of the lock, thus opening the door. When the guest finally leaves the room, he does not hand in the card but is invited to keep it; the lock is automatically scrambled and a completely new card issued for the new guest. The card now in possession of the old guest is useless and will not open a door in that hotel again. Whilst installing these systems can be expensive, the cost can soon be recouped through the added security obtained and the absence of the expense of changing locks and cutting new keys, which is now a costly operation itself. Details of these systems are available from the manufacturers or from the local police crime prevention officer.

Windows

The security of windows in rooms has already been discussed in chapter 3 and need not be covered here, except to restate the importance of windows in room security. Window security is especially important where the room has a balcony, as often many rooms can be reached from one balcony – another reason for even empty rooms to be secured properly.

Hotel property

One factor not to be overlooked is the security of hotel property in the rooms, which is often stolen. Towels seem a particularly attractive item. Hotel property is always marked, but this does not seem to prevent the theft. One hotel hit on the idea of offering for sale at a cheap price a towel which said 'Stolen from X Hotel'. It apparently sold like hot cakes. It is possible to have devices fitted to larger items, such as television sets and refrigerators, which will activate as soon as they are removed, but this type of system needs to be considered in the building stage of a hotel and not afterwards because of the wiring needed. What an existing hotel can do is to ensure that any identifiable property is recorded and any serial marks etc., noted for future reference.

Security patrols

The final security measure necessary for rooms is the security patrol of the corridors, which will be carried out by security officers and will be a great deterrent against the room thief. Part of the patrol will be an examination of all room doors and any keys left in the doors should be seized by the patrolling officer. He should first check, if there is someone in the room, that they are the bona fide guest; if they are he should hand them the key and point out the dangers of leaving it in the door. If the room is empty, then he should keep the key until it is claimed. He should also ensure that rooms are kept shut during cleaning. Hotel staff can also play their part in this type of security, particularly room maids, who are more likely to see something suspicious than any other member of staff. An observant, aware hotel staff are the best front line defence against attack on rooms.

Investigation of losses from rooms

Whenever a loss is reported it should be investigated immediately, by the security officer if there is one, or by the duty manager if not. The level or seriousness of the loss will determine whether or not the police are involved at this stage. The investigating officer will need the following information:

1 Full name and address of the claimant
2 A full description of the missing items and any identification
3 When the item was last seen by the claimant and when the loss was actually discovered
4 Whether the claimant had any visitors; whether he or she is suspicious of anyone in particular – if so why
5 The value of the missing items
6 Details of any insurance held

This information should always be taken in writing and when all possible information has been obtained the claimant should be invited to sign the document. The hotel should prepare a pro forma for such losses listing all the relevant information required. Such a document can act as a good *aide-mémoire* to the investigating officer. If the claimant refuses to sign any such claim the fact should be recorded as it may be relevant; it has not been unknown for false claims to be made.

When the claim is complete the investigating officer

should examine the guest's room, having first out of courtesy asked to be allowed to do so. It is not necessary for the claimant to be in the room at this time, but in some cases it might be considered advisable. The main points that the investigator should be looking for will be:

1 Is there any sign of forced entry to the room?
2 Are all the locks and other devices in proper working order?
3 Are the missing items definitely missing? A search must be made as items often are found in other places.
4 Were there other valuable items in the room that were untouched?
5 Was anything in the room tampered with? e.g. suitcases, briefcases?

The investigator should then be in a position to commence his investigation. His starting point will be with the staff on duty, in particular the room maid. Did the maid ever see the missing items in the room? Did she see anyone entering or leaving the room? Did the guest make any comment to her about the loss? The enquiries should then extend to other members of staff, such as valets, porters etc., who may have been on the floors. Particular attention should be paid to floor service waiters, who will generally speaking have a master key. Enquiries might also be made of the other guests on that floor, but this will depend upon the seriousness of the loss and their availability. The last thing the hotel will wish to do is to tell other guests that there has been a loss from a room.

If the loss is a large or serious one the hotel's insurance company must be informed at the earliest opportunity, as must the police. This step is important where the theft is of jewellery or large sums of foreign currency, as these have to be disposed of, and a circulation to likely places at the earliest opportunity is imperative. At no time should any admission of liability be made to the guest by the security officer or any member of hotel staff. Any subsequent claim will be dealt with by the insurance company who will work within the framework of the relevant law and will only meet large claims where it can be proved that the hotel admit liability or there has been negligence by the hotel staff.

The importance of the initial interview with the claimant cannot be stressed enough, as it often reveals the true facts. For example, a guest who has been to a nearby gaming club and lost a large sum of money does not wish to say so, either to his wife when he gets home, or to his firm whose money it may be, so he may invent a room theft. In the same way neither may a man who has lost money following his involvement with a young lady of dubious virtue wish to report the fact. Certainly such possibilities should be borne in mind whenever large sums of money are involved. In the same way, where items of valuable jewellery are reported stolen, a quite natural question to the claimant could be 'Why did you not take advantage of the hotel's safe deposit as recommended in the leaflet issued with your room key?'

The investigating officer has a duty to the hotel as well as to the guest to discover the truth and should conduct his enquiries to that end. If his enquiries indicate a suspect for the offence then he should report his suspicions at once to the police, who will then take over the enquiry. It is far better to handle the case in this way, since if the investigating officer starts interrogation himself he may well, with his lack of experience in the field, spoil the subsequent investigation. Also he should remember that the suspect might be responsible for other offences.

Should his enquiries lead nowhere, then the offence should be recorded and some research done on similar offences in the past. This type of investigation can often have good results. A number of questions can be asked:

1 Has this room been attacked before?
2 Has the key of this room been reported as lost?
3 When was the last similar offence reported?
4 Who were the staff on duty at that time?
5 Has this guest reported a theft before?
6 Have the times of the offences any significance?

These few questions can often suggest another line of enquiry, but these questions can only be answered if proper records are kept. Under no circumstances should notes just be made on scraps of paper, but the record of the theft should be made on a properly prepared form and then recorded in another book or filed away in an easily accessible way.

Investigation of room losses should be treated not just as another chore but as an important task; in that way not only may the offence, if any, be cleared up, but the detection will act as a deterrent in the future.

7   Security of property

Under hotel legislation liability is limited unless negligence can be proved or the property is deposited with the proprietor for safe keeping and is lost. As negligence is easy to allege but not so easy to refute it is most important that hotels take all reasonable steps to safeguard the property, money and valuables of their clients, and are seen to be so doing.

Limiting legal liability

The UK Hotel Proprietors Act requires the proprietor to have a copy of the notice set out in the Schedule to the Act conspicuously displayed in a place where it can be read by his guests so that he can take advantage of the limit of liability under the Act. This he will obviously want to do and he should also make ample provision for the security of money and valuables belonging to his guests, so that they are given the opportunity to put them into safe keeping which will also limit his liability. It is those measures that will now be discussed.

The proprietor should make quite clear to the guests, in the literature issued to them at the time of their registration (either in the form of a small booklet or a printed card), what facilities he has available for the security of money and valuables. The guest should be entreated in the literature to take advantage of these facilities and should be advised that if he does not take advantage of these measures and has valuables stolen, the hotel owner's liability will be limited and the guest may have no redress at all. The reception staff should always advise guests orally to read the card or the notice, and then the hotel owner will be in a position to prove that he has done everything in his power to bring the position regarding security of valuables to the attention of the prospective guest.

Facilities for safe deposit

The level of security that he needs to supply will depend, as with most other security, on the size of the hotel and the number of rooms available for let. In addition, the type of business of the hotel will be important, as a hotel at the top end of the market may well have to make more measures available than one at the bottom end.

The theft or loss of cash must be one of the main hazards facing the hotel, and cash is, of course, one of the easiest items to dispose of without fear of detection. If money is stolen from a room it is most unlikely that the guest will be able to give any information or details, or indeed be able to provide any identification at all, and without some definite line of enquiry the chances of recovering that money will be very slight.

Safe deposit boxes

The best available protection for money and valuables, such as jewellery, is the safe deposit system. For this system to be effective very clear guidelines must be laid down, and the whole system effectively supervised. Each security box should be numbered and should be locked with two keys, one of which will be held by the guest and should only be handed over to a guest of the hotel who is registered at that time. This point must be made, as regular users of a hotel are rather inclined to expect that the safe deposit facilities will still be available to them even when they are not in residence. Use of the facility when the guest is not resident should never be permitted, however important the guest, as it reduces the number of boxes available and thus renders the hotel less able to safeguard property. The second key to the security safe box should be held by reception, preferably in another small safe. It will always be necessary to have both keys to open the box, therefore both the guest and the member of hotel staff need to be present at the time of the opening.

The safe deposit boxes must always be sited where they can be supervised by a member of staff, never in such a position that they will be unattended for long periods. The supervision of the system is imperative, and should entail a daily log of all safe deposits in use. The log should be handed over from shift to shift by the reception or cashier staff responsible, and should be signed by both members of staff as correct. At least once a day this log should be checked by the duty manager and/or the duty security officer, who should check that the correct handing over procedure has been followed, and that the details shown on the log are correct. He will have to check each safe shown as empty on the sheet against the keys held, as for every safe that is empty and out of use there should be a corresponding key available for use. If there is not, then immediate enquiries must be made. Periodic checks should also be made to ensure that the persons shown on the log as using the safety deposit box are, in fact, guests of the hotel. If they are not, again enquiries should be made. The log should be signed by the inspecting officer on each occasion. All these measures will show, should there be subsequent problems, that the hotel has done everything within its power to see that proper control is maintained.

One of the major problems to arise with the safe deposit system is the guest who has lost the key. The question of lost keys to safe deposits should never be dealt with by a member of the reception or cashier staff, but always by a security officer if present or the duty manager. The original report of the lost key will be made to reception or cashiers and must be notified immediately to one of the two persons nominated above. They should question the guest very carefully as to the circumstances of the loss, whilst at the same time ensuring that the person making the claim is in fact the person entitled to hold the key. Some evidence of identification will nearly always be necessary. Under no circumstances should a spare key be issued until complete satisfaction as to the bona fides has been established. If there is any doubt at all no key should be issued.

It is always advisable as a matter of policy not to inform guests that spare keys are available, but it should be made clear to them at the time of the issue of the key that if the key is lost a set fee will be charged and they will have to wait for another key to be cut. Such a prospect will tend to make them extremely careful, and less likely to lose the key. The charge should be more than nominal, as eventually the lock will need to be changed and a further key cut, which will be fairly costly in itself. It would be advisable, in fact, for hotels to keep spare keys for all safe deposit boxes, stored in sealed packets, i.e. a packet with the key sealed inside with a lead seal that must be broken before the key can be removed. These spare keys should be kept in a locked cabinet, preferably in a large safe, which should be situated in the security office if there is one, or in the general manager's office.

When a key is reported lost, and after enquiries a

spare key is issued, then it is strongly advised that, when that particular guest has finished with the box, the lock for the guest's key should be changed. The changing of locks should be handled in the same way as discussed in chapter 6: a small number of spare locks should be held by security to be fitted in these circumstances. Then the lock that has been removed can be taken to a locksmith for a second key to be cut. In this way each safe deposit box will always have two keys available for use.

Safe deposit bags

In a small hotel it will no doubt be thought unnecessary to go to the expense of installing a safe deposit system, but the hotel must still make some facilities available for the protection of money and valuables. One method would be to have a system of deposit envelopes which can be sealed and signed by the person making the deposit; a ticket is then made out in duplicate for each bag, and one copy is issued to the depositor. The bag is then deposited in a safe, the key to which is held by a nominated member of staff, with a duplicate key in another safe either in security or the manager's office; this second safe should have only limited access. Should a deposit bag ticket be lost by the guest, then again this loss should be investiaged until the hotel are satisfied that they have a genuine ticket holder.

Large items

There will be many valuable articles that a guest may wish to deposit, such as fur coats, cameras etc., that will obviously not go into the normal size safe deposit. For these items a large fire resistant safe should be available, situated away from the reception area; it is normal practice to have this safe situated in the security office. The articles should be taken personally by the security officer or, in his absence, the duty manager, a printed receipt form should be issued showing full details of the item deposited, together with the personal details of the guest, and the deposit should be recorded in a separate register. Such deposits must then only be handed over against the production of the receipt, and against signature. Should this receipt be lost very careful enquiries will need to be pursued before any property is handed over. It may be necessary to bring back the member of staff who received the item to identify the depositor.

Very occasionally extremely large items will be deposited; one deposit within the author's knowledge was all the instruments of an international orchestra. In such circumstances the only course open to the hotel is to store all the items in one room, which is then secured by double locking or some other means. It may be considered necessary to add other measures such as padlocks to the door, and if possible the room chosen should not have windows. An important point with such deposits is that two members of the hotel staff should be present when the items are handed in and insist that each item is checked in individually. This may take time, but it will safeguard the hotel against any allegations made later.

Lost and found property

The security of any property left by the guest in his or her room will normally be the guest's own responsibility providing that the hotel has taken all necessary steps, including proper procedures by staff, such as always ensuring that rooms are shut when cleaning etc. is carried out. However, staff should be trained to assist in general security by bringing to the attention of the management any property they consider valuable which is left lying about in rooms and is thus liable to lead to temptation. This latter point is one of the reasons why a firm policy should be laid down in respect of lost and found property.

Lost property

Lost property does not create too many problems, the main factor being to ensure that the loss is properly recorded and that full details are taken. A rough description will not be enough, as often property comes to light after the guest has left the hotel, and without enough details positive identification proves difficult, especially where the guest is from abroad and has returned to his own country. Again the recording of details is helped by a form that can be filled in at the time of the reported loss, and the forms should be filed in an accessible place. The person taking the report of the loss should always be satisfied that it is a loss being reported, not a theft.

There will be many occasions when property alleged to have been lost will not have been found even after some time. If this is the case then the guest should be so informed. The fact that an item reported lost in the hotel has not been found does not mean that it has been stolen and the procedure for dealing with theft should not be commenced unless there are suspicious circumstances surrounding the loss. Guests do not spend all their time in the hotel and property can be lost anywhere.

Found property

Found property presents more problems, and the amount of property left in hotels is surprising. The object of any system to deal with found property is to ensure that any such property is handed in. Many items of property left behind are very attractive, and could tempt the member of staff finding them to keep them. Firstly there must be a strict rule that all property found must be handed in to the member of staff's department head as soon as possible. At that time it should be properly recorded with the time, date and place of the finding being stated. There should then be one central point for the handling of all found property, which should be the security department if there is one. This found property should be collected on a daily basis, and again recorded in the central handling point. It should then be properly labelled and locked away. Items of particular value and money should always be locked in a safe and not placed with the other property.

A level of value should be fixed with property, from which a system of returning the property to the finder can be drawn up. It is recommended that any items to the value of £25 or less be returned to the finder if they are not claimed within three months. Items of over this value should be kept for a period of six months and then also handed to the finder, who should be invited to sign a disclaimer accepting responsibility for the property as against the hotel. The benefit of this system against other systems such as selling the property or disposing of it by auction to the staff is that is encourages the staff to hand in the property, as they are secure in the knowledge that, if it is not claimed, they will receive it back in due course.

Property that is not wanted by the finder, or was found by a member of staff who has since left, can be disposed of in whatever way is decided. Small valueless items should be destroyed.

There are, of course, certain items that should not be returned to the finder; for example, spectacles and medicines or any drugs. Drugs, etc., should be destroyed.

In a small hotel it is possible to arrange to return such property to the loser where it is found in a room after the guest leaves, but in a large hotel, where perhaps the room changes occupancy every day, this system is not practical, and in such cases the hotel will await a claim from the loser and not waste time contacting a large number of guests in order to discover the owner of the property. When a claim for property is received, the records should be checked at the central point, and if the property has been found and can be identified beyond doubt then it should be returned. If there is any doubt, the property should not be returned until further enquiry has been made of the claimant.

8 Control of food and drink

Although the control of bars, restaurants and floor service is a management problem, security must be involved in the setting up of the procedures, because it is in these areas that offences will be committed by staff, resulting in an economic loss. The experience of management and security staff in detecting offences in the past can be put to good use in setting up systems which will prevent similar offences in the future and thereby minimise losses.

Any hotel management that believes it has succeeded in keeping out all undesirables and has kept all its employees honest is living in cloud cuckoo land. In the same way no manager should accept that thefts are inevitable and part of the business, as they need not be so. A few items missing each day will not amount to a great deal of money, but added together over the period of a year the total amount will give an entirely different picture. The whole point about such losses is that they affect net profit. Every pound that is stolen or fiddled has to be replaced, and to replace this pound will need much more than a pound in additional business.

The need for security systems

In other parts of the book we have discussed the need for physical measures such as locks and safes, but just as essential are the measures to protect hotel materials such as food and drink; in other words, the documentation necessary to control the related activities. If some effective method is not established losses may go unnoticed and can get out of hand.

The documentation system needs to be checked and supervised in the same way as any other security measure, as does the control of the documentation. All managers and security officers should be aware of this control system, know how it works and be able to check that it is working. Too often the people responsible are not conversant with the systems, which can enable a fraud to start up and go undetected sometimes for years with a resulting heavy loss.

At this stage one example will suffice. It concerns a fraud perpetrated by floor service waiters in collusion with the floor service reception. At the hotel in question it was the rule, as indeed it should be, that no floor service waiter should receive cash payment for meals, etc. delivered to rooms. As far as was known the rule was adhered to until one morning two guests queried their bill on booking out. It transpired that they had obtained a meal during the small hours of the morning, had paid cash for it but had never received the change. It was the change they were questioning. The records held in the control office were checked but no trace was found of any meal delivered to that room. The cancelled tickets for that evening were then checked and a ticket in respect of the meal was found, but it had been marked as order cancelled and filed in that way.

If the hotel system had been strictly observed this could not have happened. The system was that the girl on floor service reception took the order over the phone on a ticket in duplicate. The second copy of the ticket would go to the kitchen for the making up of the order. After the order was completed this duplicate ticket was placed on a spike and the meal collected and taken to the guest with the top copy of the ticket, which he should sign so that it could be placed on his bill. The duplicate tickets would be removed from the spike in the kitchen by the night chef at about 4 am and placed in a metal box in the head chef's office. This box was then secured by a padlock of which only the chef had the key. So on the face of the known circumstances it should not have been possible for the guests to have obtained a meal. However, the system was not being adhered to, and it is unlikely that any existing member of management or security was aware of what the system should be. In fact, for at least four years there had been no padlock on the box in the chef's office. After the night chef went off duty the girl on floor service reception would go to the box and remove the duplicate copies of tickets for which she knew cash had been received. She would then staple these tickets to their top copies and mark them cancelled.

One place where this fraud could have been detected was the control office for the tickets, but again insufficient instruction had been given to the staff. The only instruction given to the man responsible was to check that all tickets had been accounted for, and that any cancelled tickets had both copies. At no stage had he been asked to report if an unusual number of cancelled tickets had been submitted, nor had he been asked to examine the cancelled tickets: if he had been he would have seen that the bottom copies had small holes in from the kitchen spike, and in many cases gravy stains!

It transpired, after questioning, that this fraud had been going on for some years, so the amount of loss to the hotel was probably enormous. The whole fraud could have been prevented, and if not prevented, it should have been detected in the very early stages by proper training, supervision and knowledge on the part of management and security. This story, it is hoped, will illustrate the type of fraud that can occur and in addition show that proper systems must be instituted and supervised.

Bars

One of the high risk areas and targets for theft is the stock of alcohol, because it is an attractive item and can be easily disposed of. The cash side of alcoholic sales is always open to abuse by the unscrupulous member of staff.

Normal practice in a hotel is for the sale of alcohol to be under the control of the food and beverages department, with a manager as the department head. He will be responsible for ordering the alcohol in bulk from the wholesaler, and the first check will be to ensure that the alcohol ordered is actually received. The hotel staff must be prepared to check in all goods ordered item by item, and it is equally important where sales of beer are concerned that empty bottles, cases and kegs are also checked out and that the number collected is correctly recorded. Any breakages or damaged goods must be noted and accounted for. Following the delivery the goods must be stored in a suitable locked storeroom and again it is advisable to check each item as it is stored.

Bars are normally stocked with a predetermined quantity of each item before each set of opening hours. The head barman or other person authorised should gather empties, etc., at the end of each tour, and make out a requisition for the replacement stock required. This requisition should be filled from the store and then transported to the bar where the goods should again be checked in item by item. If the requisition is in order it should be signed as correct by the person receiving.

In this way the stock kept in a bar is guaranteed correct and deliveries to and from a bar can be properly recorded. In consequence the reconciliation of the stock to sales from the bar, the next step, is easier. Never a simple task, it will always be assisted by the knowledge that any reconciliation commences from a secure base.

The amount of alcohol that can be dispensed from a bottle is known, and each hotel must fix its own allowances either side of this; for example, the number of tots to be obtained from a bottle of whisky might be fixed at between 30 and 32. The same applies to kegs and barrels of beer, as certain allowances must be made for the cleaning of pipes etc. Once the allowances are known, it is a fairly straightforward task to compare the actual sales made with the potential sales from the known stock. They will not, of course, tally exactly, but should be within reasonable limits of tolerance. If those limits of tolerance are exceeded one way or the other an immediate investigation is called for.

The above is not, however, the whole story, as there are other ways for an unscrupulous barman to make money which will not show up in any stock take or reconciliation. One will be from the practice known as 'buying out', where the barman goes to the nearest discount liquor store and buys certain stock which he places in the spirit dispenser himself, taking the whole of the proceeds. If he is careful enough to buy the normal hotel brand this is very difficult to detect. Effective management can overcome the practice with proper supervision and keen observation in the bars. Visits to the bar at peak time will soon ascertain the level of trade being taken. If this level is not reflected in the bar takings then investigation is necessary. Another method of detecting 'buying out' is in the orders received in the stock room for mixers, that is tonics, dry gingers etc., usually served with spirits. If the consumption of these is far in excess of the bottles of spirits supplied, then again suspicion should be aroused.

It would never be possible to outline in any book the many ways bar staff can use to steal from or defraud the hotel. The truth is that every day someone finds a new method of crime, and not just the staff. The best way to overcome crime is to have a set policy concerning bar staff, which could include many factors. These will depend upon the activity of the hotel and cannot be set out in any general form, but they could include the following:

1 The previous employment record of all bar staff and other staff handling alcohol should be very carefully checked prior to employment. If there is a large bar operation the fidelity bonding of employees could be considered.
2 The hotel should ensure that all glasses, optics and other measures comply with the various regulations in weights and measures so that no opportunity is given for the excuse that staff were unable to give the right measure.
3 Bar staff should not be permitted to serve 'free' drinks or 'stand a round' under any circumstances. Any entertaining should be done by a member of management and the drink consumed accounted for as with any other sale.
4 Bar tills should be checked at frequent intervals by management, and where there is a change of shift, each barperson should have his or her own till drawer, which will enable a tight check on individuals to be kept.

As a deterrent against collusion with other members of staff, which even the most stringent security and management measures will not detect, an outside agency may be employed to make test purchases and generally keep observation on bars. The personnel making these visits will be unknown to the bar staff, as will the times of visits, and the mere knowledge that a watch is being kept is one of the best deterrents.

Food and beverages

Under this heading comes the control of food and drink,

drink meaning in this case, tea, coffee, soft drinks, etc., and, to a certain extent, wine. Food is again an attractive target for the thief and it is tempting to remove small items, especially items which are highly priced in the shops. Generally speaking the theft of food by hotel staff is a haphazard practice, but it could add up to a considerable amount over a period and must, therefore, be controlled.

Deliveries. As with alcohol, the first line of defence against theft and fraud will be the effective checking of deliveries. Goods should be delivered to one place only and the staff handling the deliveries should be regular staff. Each delivery must be carefully checked against any delivery note, and management must be aware of the possibility of collusion between the delivery man and the member of staff receiving. The common practice for this type of theft is for the delivery man to make a short delivery, signed as correct by the person receiving, the surplus goods then being sold at a shared profit to each party. To guard against this spot checks of deliveries should be carried out by management or by security officers, if they are employed.

Issue to kitchen. Careful checks of issue from the central stores to the kitchen are also necessary; the goods delivered to a kitchen should be placed in a locked cupboard and it is advisable to ensure that refrigerators and freezers are also kept locked when not in use with the keys held only by senior members of the kitchen staff. There should be a firm rule that waiting staff from the restaurants will never have access to the kitchens beyond the serving hatch areas.

Food issued from the kitchen to waiting staff should only be given out on the production of a docket from the restaurant or area concerned, and nothing, not even a pot of tea, should be handed over just on the oral authority of a member of staff. To prevent the type of theft illustrated earlier in this chapter, all duplicate copies of food orders received in the kitchen must be placed immediately in a locked container with the key held by senior chefs only. No other person should have access to this box, and it will be cleared by a member of management.

Documentation. All orders taken, whether in the restaurants, on floor service or on lounge-type service, must always be recorded on duplicate slips. A guest at the hotel must always be asked to sign the top copy when he has completed his meal, and if cash is taken from a more casual guest then the top copy of the docket must be handed over as a receipt. This latter point needs to be carefully monitored, and could form part of another exercise for the outside agency. Another case will serve as an illustration of a different type of theft.

A lounge service waiter got into the habit of keeping the top copies of small orders for pots of tea, coffee, etc. in his pocket and not handing them over as a receipt to guests paying cash. He soon realised that people are often careless about asking for receipts. The next guest who asked for a receipt for cash would be given one of the top dockets from the waiter's pocket and he would pocket the cash. There was no real check on amounts of tea and coffee, etc., served, as that area had its own facilities for making them without going to the kitchen. The waiter was eventually caught by deliberate visits made by staff members from another hotel. He was arrested and charged by police, and the takings in his area shot up by £140 per week! Even a small fraud can amount to a considerable sum of money over a year.

The checking of all documentation must be a very careful process. Firstly, documents must not be issued haphazardly and no unnumbered documentation should be given out. For all documents issued a record should be kept giving details of the person and department to whom issued, the date of the issue and the serial number. Control must be such that all dockets can be accounted for. Where mistakes are made both copies of the incorrect document must be retained. A random check on such documentation should be made each day, and some of the cancelled dockets should be checked for authenticity.

Customer fraud. Another hazard of the food service departments, in the restaurant operation, is fraud by customers, the most common being consuming a meal with no intention of paying for it, a criminal offence which can be dealt with by police. It is difficult to prevent, and often the perpetrator will be a most unlikely individual. The staff can be trained, however, to look out for the person who does not look as though he has the means to pay. A particular danger time is late at night, especially after licensing hours, and parties of young people in obviously alcoholic high spirits should put the restaurant staff on their guard. It is always advisable for any hotel with public restaurants to close the majority of these to the general public by late evening and certainly before the local public houses turn out.

Staff catering. The final operation to be considered in this section is the staff restaurant. Control of the staff restaurant can sometimes be very slack owing to the misconceived idea that very little can go wrong. However, this operation certainly needs management attention. The amount of food issued should be checked against the meals consumed, which sometimes gives surprising results. Complaints from staff members about the quality and quantity of food should not just be discounted, but should be treated with some seriousness. It should be generally understood that meals will only be served to those members of staff authorised to be fed. With a large number of staff this is not easy, and it may require the issue of meal tickets or some other form of identification by the various department heads. If care is not taken staff off duty will come in for meals, and the cost of this can be high. It has even not been unknown for a complete outsider to gain access to a staff canteen and quite happily consume a meal without detection thanks to slack control.

Banqueting

In most hotels banqueting will be part of the food and beverage department, but in a large hotel it will be a separate operation with its own manager. The staff employed on this operation need to be very carefully selected and supervised. It will often be necessary to employ a considerable number of casual staff, particularly waitresses, and at these times maximum control needs to be exercised. Because of the nature of the operation the staff will have to have considerable flexibility on ordering of food etc., and therefore the wastage will of necessity be higher. Control needs to be established in the area of food and wine, as dishonest staff will ensure that there is a good surplus of this, which can be disposed of afterwards to their profit. There will often be a small bar operation with a banqueting function, and the transfer of alcohol to the bar gives openings for theft if not properly supervised. All alcohol issued for a banqueting function should be signed in and out of the stores and, should an excessive amount be consumed in relation to the number of people attending the function, investigation is called for.

Wines

Wines have been given a separate heading because their consumption can apply to any part of the hotel, ranging from guest's rooms to restaurants. The handling of wines can be open to abuse, not least in the type of security used to look after the product itself.

Display. It is a common practice in many restaurants to have the wines prominently displayed, which is fine for customer relations, but needs to be carefully considered from the viewpoint of theft. If wines are displayed they should be away from any public entrances, but not tucked away in an unused corner. Too often the wine rack is found in an area where a thief needs only a few seconds to take a bottle and conceal it with very little fear of detection. This is also a problem in the small hotel where the wines need to be accessible but do not have staff about at all times. One hotel tried to solve the problem by placing the wines behind the receptionist in a metal cage. This looked very effective on the face of it, but a security adviser discovered on inspection that bottles could easily be removed by reaching through the cage and pulling them out.

Large orders. Of particular note is the larger order of wine, perhaps at a banqueting function where two or three dozen bottles may be ordered. By careful husbanding not all the bottles are used, which should be of benefit to the guest but is often to the benefit of the staff. Close supervision of the pouring of the wine is necessary, and a senior member of staff present should take personal charge of the bottles.

A simple check for management at the end of any large function is a count of the empties, and a look at the labels. Substitution has not been unknown. It should be unnecessary to say that the empty tally should agree with the full issue, but this is a simple matter often overlooked.

Room service

Earlier in this chapter a story centred on the operations of floor service illustrated the vulnerability of this operation to abuse unless strict guidelines are laid down.

Documentation. It must be a firm ruling that under no circumstances will floor service waiters receive cash from guests, or ask for cash. If the guest offers cash in payment for a meal ordered in the room, the cash should be refused and the matter referred to the duty manager if the guest wishes to complain. It is recommended that the system should work in the following way. Orders should be received in one central point, and taken down in duplicate. One copy of the document should go to the kitchen for preparation of the order and the second remain at the order point. When the kitchen have made up the order the kitchen copy should be filed away in a lockable container. The order should then be taken to the guest and at all times a signature obtained from the guest on the top copy of the form. As an added check the documents could be printed in three copies, so that a copy can always be left with the guest.

All the signed top copies must be collected at frequent intervals by the duty manager who should also check on other orders received and not yet filled. These documents should then be processed quickly so that the cost can be added to the bill. The services of an outside agency can again be used as a further check. This check would be carried out by booking a room, then placing an order with room service and giving a full report on the procedure followed by the waiter.

Access to rooms. A difficult problem with floor service is the issue of master keys. Firstly a decision has to be made as to whether or not the waiter receives a master key. If the decision is not to issue a master, part of the service offered will be lost, as guests will be inconvenienced to a certain extent by having to answer the door, perhaps from bed or the bath. Against that is the fact that the master key issued would have to be a key allowing entry to all rooms, not just a floor master, which means yet another member of staff having a master key, another security risk. The policy on master keys can only be decided by senior management, and will depend upon the level of service required and to a great extent the size of the hotel. If such a key is issued then it must be signed for and signed back at the end of the tour of duty. A further problem can arise with the collection of the tray and dishes after consumption of the meal. On balance it is better if such utensils are either left outside the room door by the guest or removed by the room maid during cleaning. It is not advisable to have the room service waiters re-entering the room to collect them. The reason for this is quite simple. When the food is delivered, the guest will normally be present. When the empty tray is collected he may not be, and it is not advisable for any more staff than necessary to enter occupied but empty rooms. Such a policy is also a good security and supervisory measure, as any waiter entering a room other than with an order can be immediately questioned and invited to give an explanation.

All the operations described in this chapter have security problems, and all of them are open to abuse by staff if proper controls are not instituted. This abuse must be controlled, often because the losses that ensue are completely unknown to the hotel. The two examples quoted in this chapter had both gone undetected and only after the offences came to light did the management realise the considerable loss they had been suffering.

The secret of any good management and security is that proper checks and effective supervision are not only done but seen to be done. Paying lip service to security and supervision is not enough. The majority of hotel staff are honest, hardworking people who just want to earn a living, but there will always be a minority looking for ways of supplementing their earnings. These people may also be honest in the beginning, but will not remain so if the systems employed by the hotel are so lax as to allow them to defraud without much chance of detection. A system like this will not take staff long to discover. They will look at the systems, weigh up how effective they are, whether or not they are used properly and if they are checked. If their investigation shows obvious weaknesses, then management beware.

9 The housekeeping department

While a guest may occasionally complain about the quality of food, the service in a restaurant, or the tardiness of a cashier, it is certain that the department that will attract the main complaints will be housekeeping.

Everything to do with the guest's comfort while at the hotel will be based on his room, and should his bed not be made, his linen not be changed, his laundry go missing, his shoes be uncleaned, or some item disappear from his room, then housekeeping will be the object of his wrath. It is also true, but usually goes unnoticed by the guest, that an efficient housekeeping department will mean an excellent stay, and a good recommendation for the hotel. But how many guests tip the room maid? Equally true is the fact, often stated in this book, that an alert, observant housekeeping staff are one of the best security measures open to a hotel, and a security officer will neglect his liaison with housekeeping at his peril.

Whilst the main duty of housekeeping is the care of the sleeping accommodation, it has many other functions, which will be controlled by the head housekeeper, including:

house porters
cleaners
laundry
valet service
linen

Room cleaning

Without doubt the most important aspect of housekeeping is the cleanliness of the guest's rooms.

Chambermaids

Access. It is usual practice in a hotel for the care and cleanliness of each room to be entrusted to one room maid or chambermaid. She will have a number of rooms to clean and will usually be supervised by a floor housekeeper. The nature of the job means that the maid must be given unlimited access to the rooms on her section, unless they have been double locked for security reasons, so she will have more opportunity to steal from the room than any other member of staff. No one will query the presence of a maid in a room and she will have ample time to make a thorough search. These facts alone should cause management to make every effort to ensure that all personnel employed on this duty are honest, which means that thorough pre-employment checks are essential.

The problems concerned with the issue of master keys were discussed in chapter 6. The key usually issued to maids is a floor master and for this operation consideration could be given to dividing a floor into sections. However, more problems might be created than solved, as it is not always possible to allocate a particular set of rooms to a particular maid. Each maid should sign for her key before commencing duty and should return the key immediately the duty is finished. These keys will normally be handled by the housekeeping department.

Training. All maids should be given training when they

commence duty at the hotel. The training will obviously include instruction on how to carry out their duties, but should also include the initial fire training required by law and a session on security, given by the chief security officer if there is one. The security aspect of the duty should be stressed and the maid should be instructed on all relevant points.

Keeping doors shut. When working in a room they should keep the door shut, but they should be issued with a suitable notice such as 'Maid working in room' to put on the outside of the door. The object of the notice is to warn the returning guest that the maid is in the room, and to assist the floor housekeeper or whoever is supervising to find the maid. Another school of thought on this subject suggests that it is better for the maid to work in the room with the door open. It is said that the open door enables the maid to observe what is going on outside, but it is most unlikely she will do so, as maids are hard working and once in a room are too immersed in what they are doing to see much else.

As room cleaning is hard work, often a maid will try to cut down her work by cleaning more than one room at once, and to this end will leave the door of one room propped open with her trolley or perhaps the vacuum cleaner while working. This practice must be eradicated by the supervising staff. Doors propped open create a very great security hazard, and are an invitation to the sneak thief to steal.

Often a maid will be approached whilst carrying out her duties by a guest who has mislaid his key, and will be asked to assist by opening his door with the master key. Under no circumstances should the maid do this, but she must send for the nearest housekeeper. This will obviously annoy the genuine guest, but if the housekeeper explains why it is done the guest will understand and will appreciate the maid's position. This rule must be adhered to, as many thieves operate by pretending to be guests, and the hotel will have to accept the full responsibility for any loss incurred under these circumstances.

Reporting suspicions. Where a room theft occurs, obviously the guest, if not the hotel management, will suspect the maid first. All maids should be advised of this fact, and that they are the best persons to protect themselves by reporting anything that they believe to be suspicious or out of the ordinary. To ensure that they do so hotel management must inform the maids of everything going on on their floor, so that they are fully aware of anything that might affect their jobs. If a maid is kept fully aware, she will return the confidence in her by reporting anything unusual or suspicious. In many cases maids have discovered and reported not only thefts but fires, and they have also been instrumental in catching room thieves.

The maid should also be advised to report suspicious circumstances concerning the room itself; for example, if a single room is occupied by two people. Many potential 'walk outs' have been prevented by alert maids who have reported that there was scarcely any property or clothing in the room and that the guest appeared to have very little to support the stay.

Another of the maid's duties which can be very helpful is to record the presence of other members of hotel staff who enter a room whilst she is there, such as floor service or valets. One case of room thefts was solved after a maid reported that a floor service waiter had come into the room she was cleaning. A check revealed that no order for floor service had been placed from that room and subsequent enquiries brought an admission of other rooms thefts.

Floor housekeepers

A very important part of all these duties will be the level of supervision by hotel management. Supervision will normally be in the hands of the floor housekeeper, who will have a master key, so again effective pre-employment checks on the individual holding this post are essential. Floor housekeepers should be on the floor at all times and will have four main tasks:

to check on the quality of work carried out;
to discuss any particular problem concerned with the room;
to pass on any instructions;
to carry out a security function.

This last function means acting as a secondary security patrol, to ensure that only guests obtain legitimate access to the rooms and to carry out security checks on maids' trolleys and her working area, particularly the linen closet. One series of thefts was detected when the floor housekeeper found property hidden beneath sheets in the linen closet. She will also be alert for the suspicious person seen on the floors, and will often be the first person at the scene of a room theft.

Other housekeeping staff and services

House porters

House porters are normally male staff who carry out the more manual tasks on the floors of the house, such as fetching and carrying bulk items. They are not involved with guests as a general rule and are supervised on a day-to-day basis by the housekeepers. Their work does, however, occasionally take them into rooms, and they

are always about on the floors. There are opportunities
for them to steal so they should be closely supervised.
They also should be security conscious and aware of the
various security problems and be alert to the unusual
and the suspicious. They should not be issued with any
keys, and if they require entry to a room at any time they
should be accompanied by other members of staff.

Banqueting. One of the duties with which they may
become involved is the breaking down of a banqueting
room after a large event. On this type of occasion they
may find items of property left by the people using the
room. They should be properly instructed on the
handling of such property which should be dealt with as
described in chapter 7. As they will not normally handle
property it may be necessary to issue specific instructions
at the relevant time. Such breaking down operations
should be supervised and any security officer on duty
should make a point of paying a visit.

Cleaners

This paragraph concerns the cleaning of areas other than
the guest's rooms, but under the control of
housekeeping. Cleaners carrying out this operation will
have have access to rooms and will usually be operating
during the hours when the hotel is at its quietest. Part of
the operation will cover the bars and restaurants and at
these times the supervision of these staff should be very
strict. These staff will often have trolleys, buckets and
other utensils in which it would be possible to secrete
hotel property, so spot checks on these should form part
of that supervision.

Laundry

A laundry service may or may not be one of the services offered by the hotel. If it is a service on offer then, depending on the size of the hotel it will be either carried out by the hotel itself or, much more commonly put out to contract, usually to a local laundry. The danger from a security view is in the handling of these items. The normal practice is for the guest to deposit the laundry either in the room or at a central point together with a completed list or form. This list must be checked immediately and any discrepancy raised with the guest as a matter of urgency. If this is not done at once, considerable problems can arise if items are alleged to be missing later. Once collected, the items to be washed should be stored in a central place and then handed to the laundry against signature and another check. On return from the laundry each item should be checked in and again any discrepancy should be immediately pursued. The guest's laundry should then be returned to him, personally if possible so that he can be invited to sign a receipt saying that the property has been returned in good order.

Valet service

Valet service still exists in some larger hotels and where it does it may be advisable for this service to supervise the laundry, if feasible. The normal service will provide cleaning and pressing of suits etc., and usually a part of the service entails the staff member visiting rooms personally. Valets should not, however, be issued with a key. They should either call when the guest is present or ask the floor housekeeper for access. In this operation as with the laundry great care should be taken with the checking in and out of items received.

Linen

The linen under this heading is purely the hotel linen, its purchase, storage and laundry. This subdepartment of housekeeping must not be in any way involved with the property of guests. The hotel property and its handling should be kept as separate as possible from any operation concerning the guests, so that there can be no allegations of guest's property being mislaid with that of the hotel.

New stocks. When the new stocks of linen are purchased, they are delivered in bulk and of course such deliveries must be very carefully recorded and checked. Once the delivery has been received to everyone's satisfaction the linen must be stored in a secure area under lock and key.

Identification. The next essential is a satisfactory system of marking this linen. Larger hotels will no doubt have their name already woven into the fabric, and it is advisable to have some non-removable identifying mark placed on each item before it is taken into use.

The linen room will have two basic functions: firstly the issue of new linen from stock to replace worn, damaged or lost linen, and secondly the handling of soiled linen for laundering. A considerable movement of linen each day is entailed, and an accurate record of all these movements must be maintained in the linen room. A stock of clean linen will normally be kept on each floor from which the maids will draw their daily supplies. Maids should not draw linen direct from the linen room. The stocks on the floors should be checked, daily if thought necessary, by a member of the linen room staff. At least once a year a complete stock take must be undertaken and, if the linen room records have been properly kept, this stock take will quickly show if there is a major loss. There will always be a certain level of loss from damage and wear and tear, but any substantial loss above this level should be investigated.

Loss of linen. There are two possible causes for the disappearance of linen, carelessness in handling and theft. Carelessness in handling can be controlled by proper supervision. Examples of such carelessness would be maids using hand towels to mop up messy floors and then throwing them away and the careless use of table napkins by waiters. These may seem small things to the individuals themselves but if other members of staff are similarly careless the loss can be quite extensive.

Stealing, whether, by staff or guests, is always more difficult to control. Theft by the guest will be confined to room linen, but staff theft can cover the whole field. The linen room staff should be aware of the problem and should keep a very close eye on the various linen requisitions received. It should be fairly obvious if one floor uses far more sheets and pillowcases than another, but this type of check is often overlooked. In the same way excessive requests for table cloths and napkins from restaurants should be checked. These checks will not be difficult, for if the amount of soiled or damaged linen returned does not correspond with the new order, then that new order of linen should still be in stock. If it isn't, then where is it?

The housekeeping department is, as outlined, a very important part of the hotel, if not the most important. It is also the most important from the security angle, and properly used it will help to secure the hotel in the widest sense of the word, which will do more for the good name and reputation of the hotel than any advertising.

10 Other hotel services

Some services may be offered by the hotel which have not been covered by the foregoing chapters such as the parking of guests' vehicles, pageboys or bellmen, doormen, and services connected with the hotel switchboard. These will now be covered in more detail and any security problems that are likely to arise will be highlighted.

Car parks

The main problem concerns the parking of vehicles. The provision of parking space must now be very much to the fore in the planning of any new hotel or motel and will be based on the following factors:

   the size of the planned hotel, the number of rooms and the banqueting function
   the availability of parking in the area – e.g. proximity of a controlled car park
   the local planning restrictions
   the financial consideration

If at all possible parking should be available within the premises themselves, preferably underground. Access to the parking area should be easy and clearly marked. The spaces in the parking area should be clearly defined, and if the area is large manned attendance is necessary.

Security. Having provided this facility the management must, if they are to continue to attract guests, ensure as far as possible that the security for cars using the parking area is adequate. They may well take the easy way out and post notices all over the parking area disclaiming responsibility for any theft or damage, but it is not good management to rely solely on that, nor will it encourage the guest to return if he suffers in this way. It is far better to take some reasonable steps and then post the notices.

Parking cars. Some hotels will offer the service of actually parking the car for the guest. This is a helpful facility, but management must first ask a very simple question, 'Is the person parking the cars licensed to drive?' This may seem a stupid question but it is not unknown for it to be overlooked and an assumption made that the person is a licence holder because he says he can drive. Secondly the management must establish the competence of this person: the fact that he has a licence does not necessarily mean he is a good driver. A test should be given and close supervision paid in the early stages of his employment. If these steps are not taken and he damages the guest's car and perhaps others, the cost in damages to the hotel could be large and the cost to its reputation even larger.

Car park charges. Where the hotel is in partnership with a car park company, a charge may be levied against the guest for the parking. The normal practice is for the hotel to reimburse the charge, often by the signing of the ticket issued at the car park entrance. Very clear instructions should cover this operation, as this type of transaction could give an opening for fraud to the unscrupulous car park attendant.

The entrance to the hotel from the car park must be monitored. It is not advisable to allow direct access from the car park to all hotel floors either by lift or stairway. The best course is to have a two-tier system of stairs or a lift to the ground floor, then another set of lifts and stairs to the rooms. Routing access via reception protects against entry by a thief, and makes it more difficult for the prospective 'walk out' to leave.

Security risks. Many problems can arise in a car parking area and the main ones could be said to be:

    damage to the car caused by another vehicle
    damage to the car caused through vandalism
    theft from the car
    theft of the car
    the taking of a vehicle for a 'joy ride'

These problems are not easy to prevent, particularly if the garage is an open one. Obviously the ideal situation is to have a completely enclosed garage with a secure sliding door entrance manned twenty-four hours a day by car park attendants. In most cases such control is impossible because of the expense and the hotel must, therefore, take other steps to safeguard the parking area.

Security measures. The majority of the security measures that management can take are of a deterrent nature, but can pay dividends. If security officers are employed by the hotel, then the car park area must be patrolled at frequent, irregular intervals. If no security officers are employed the area must be visited by the duty manager. The area must be well lit at night so that any movement can be clearly seen. Thieves love darkness, so light is always a good security measure. Consideration can also be given to closed circuit television. CCTV cameras should be sited to give maximum coverage and placed well out of reach so that the cameras themselves cannot be stolen! It is not always necessary to have proper operational cameras in all locations. There are very good dummies on the market which have the effect of deterring the casual thief or vandal, who will be unaware of how the camera should operate. Notices should be clearly displayed advising guests to lock their cars, and a reminder such as 'Have you locked your car?' placed in the lifts and on the stairs leading from the car park.

Damage to cars. One of the duties of the car park attendant and the security officer on patrol should be to examine all cars parked in the hotel car park. Where there is a car park attendant he should also be instructed to scrutinise all cars entering the car park for the first time for signs of damage. It is not unknown for a hotel to receive a claim for damage to a car when the damage concerned was done well before the car reached the hotel car park. If there is good evidence that the car was damaged before parking, then management will be in a far stronger position to dispute a damage claim. The regular patrols will also be able to detect any cases of damage at an early stage, so that investigation can be made. The investigation may be fruitless, but at least it shows the hotel's desire to safeguard the guests' property and that some steps are being taken to stop vandalism. All cases of damage occurring in car parks should be reported to the hotel's insurers: it would be advisable to design a form for this purpose to assist hotel staff.

Pageboys or bellmen

The duties of these personnel vary from hotel to hotel, as well as the name used. In some hotels the bellmen will carry guests' luggage and delivery packages etc., and will often be referred to as luggage porters. The name bellman is more likely in the USA. Pageboys are normally trainee hotel porters and will be used to deliver messages, page guests required on the phone, and man the porter's desk at peak times. Again these are all hotel staff who will have access to most of the areas of the hotel, and who will need to call at guests' rooms, but in spite of this they should not be issued with a master key. If they have a package, luggage or a message to deliver they should go to the room, and if the guest is out return later. Under no circumstances should they attempt to go into the room to deliver a package by asking a maid to open the door. This problem should not arise with luggage as the guest is arriving at the hotel and will be in the room. The presence of pageboys or bellmen on the floors makes them another pair of eyes for 'security' and every opportunity should be taken to stress to them this part of their duties.

Doormen

Doormen are key personnel in that a doorman is likely to be in the job for a long time, particularly at a good class hotel. The job is jealously guarded and probably one of the highest 'tipped' tasks in the hotel. As a part of security of the hotel these men can be invaluable. They get to know guests, and also, which is far more important, come to recognise the undesirables discussed in chapter 5. Properly briefed, they can prevent most undesirables from entering the hotel or, if they cannot prevent them, can notify other members of staff that they have entered.

Prevention of accidents. Instruction should also be given in the requirement of the safety aspect of the job. The entrance of a large hotel can be the scene of accidents: someone may be struck by a parking vehicle, a passenger may shut a finger in a car door, a guest may trip over a suitcase, or fall when alighting from a motor vehicle etc. In many of these accidents some liability will fall upon the hotel, so the doormen must be aware of the various problems and do all in their power to prevent them happening, certainly making sure that they themselves do not shut anyone's finger in a car door. They should not allow the pavement to be obstructed by suitcases any longer than necessary, either by moving the cases themselves to a place more out of the way or by calling for assistance from the luggage porters. Although guests' cars parked in the road outside the hotel will not be the doorman's responsibility, he should nevertheless be aware of the ownership of the cars particularly if they are parked in a restricted area and a traffic warden is taking an interest.

Hotel switchboard

Very few hotels of a moderate size do not now offer a telephone in the room. Guests can use this phone to order food and other services and to make outside calls. In the hotels at the lower end of the scale outside calls will be made by giving the number required to the girl on the switchboard, who will then obtain the call. An extra charge will be levied for this service and in order to prevent disputes at a later stage these extra charges should be very clearly set out in the room.

In the large hotel the calls will be made by the guests themselves by dialling one number on the phone, thus getting access to an outside line. This facility must be very strictly monitored to stop any abuse either by the guest or by staff. With the automatic call control this is not easy and the only effective method is to install a print out recording device, of which there are many on the market. Advice on the various types can be obtained from British Telecom in the UK or from, say, Bell Telephone Company in the USA. The normal type in use will record every call made from a room, the time it was made, the duration of the call and the number called. This print out should be checked each day without fail by the duty manager or by a security officer. He should check the calls made and satisfy himself that they have been made from occupied rooms. Many hundreds of pounds can be lost by staff making calls from unoccupied rooms to their homes in other countries, and without some recording system these calls will never be detected, particularly in the large international hotel. The print out is also used to obtain the necessary charge for the guest's bill, and as a check should a guest deny that he has made any calls. Staff who use the phones for private calls will often use a phone in a room that is occupied, and unless the guest checks his bill very carefully the staff calls will also go undetected. Management should be on guard against the guest who denies he has made a call, and should always check the print out before accepting his word. If the number called turns out to be the guest's home number, the disputed bill will be paid very promptly.

Call girls. One danger not to be overlooked is collusion between switchboard girls and call girls, but this is very difficult to control, especially at night when there may only be one or two girls on duty. A good telephone switchboard supervisor is the first essential, coupled with frequent visits to the area by security. If such collusion is suspected, the services of an outside agency should be sought.

Television sets

These days very few hotels do not have a television set

installed in the room for the convenience of guests. Far too often the guest seems to believe that the television set is included in the price of the room and takes it with him when he goes. Thefts of sets are quite prevalent but can be overcome by some basic security thinking.

Size. The first consideration is the size of the set. The smaller portable type can easily be placed in a large suitcase and carried out. The larger one cannot.

Securing. Secondly, sets can be anchored on a swivel-type fitting so that they are available for viewing from all angles but cannot be picked up and carried round the room. An expensive system is to wire each set up to an alarm which is activated if the set is moved, but this type of protection is only recommended if the theft of sets is becoming very common and economics allow.

Marking. One of the most effective security measures is a relatively recent innovation, to mark the actual screen of the set in a position where it does not interfere with viewing but is still visible to the guest. This marking is etched on the glass of the set and cannot be removed.

11 Fire

The dangers of fire

On 5 April 1974 at 0051 the fire brigade was called to a
fire in a hotel in the West of England. The fire appeared
to have started in a TV set in a communal lounge at the
front of the building on the ground floor. It is thought
that a fault developed in the set, causing local
overheating and a subsequent fire in the transformer.
From this beginning the fire spread to nearby furniture
and furnishings, then through an open door to the
staircase.

When the fire brigade arrived the fire already had a
hold and a number of people were trapped in the upper
part of the building. A start was made on rescuing these
and then the brigade found that the fire was spreading,
and that more people were trapped at the back of the
building. The rear could only be reached by a narrow
lane, so a thirty-foot extension ladder had to be
manhandled across a garden wall and between a number
of outbuildings. The fire at the rear of the building was so
intense that it was impossible for the firemen to go in and
carry out a search straight away.

It was fortunate that only two people died in this fire, a
woman and her eight-year-old son, who were found in a
room at attic level, the door to which appeared to have
been left open at the time of the fire. It was also found
that where fire resisting doors were closed they were
effective in preventing the spread of the fire. A further
eighteen people escaped from the fire, eleven of them

being rescued by the fire brigade. Nine persons including a fireman had to be taken to hospital.

This is a striking illustration of the disastrous consequences of a fire which starts in a small way but gets out of control. One of the golden rules of fire prevention was ignored, namely ensuring that electrical appliances such as TV sets are not only switched off but unplugged when not in use. As a result of this carelessness two people died.

In many countries there is effective legislation to help in the cause of fire prevention. Adherence to this legislation coupled with some simple fire prevention methods can ensure that a hotel is as safe as possible from the threat of fire, but too often the systems put in to prevent the awful tragedies that occur from fire are ignored or misused. An example of this carelessness occurred in Denmark in 1973.

The scene was an older hotel consisting of a complex of four buildings, each of five storeys and an attic, surrounding a courtyard. The construction was brick with wooden floors and a slate roof, internal partitioning being mainly of wood with some brickwork. The guest rooms, of which there were some sixty-one, were situated in the first to fourth storeys with a few in the attic storey, which also contained the staff rooms, a woodwork and paint store and a furniture store. The ground floor consisted of the main entrance, the reception and the restaurant. The entrance and reception were arranged in the original opening leading through to the inner courtyard, leaving no room for the fire brigade to get their ladders through. The main staircase was cast iron in a brick-built shaft. There were double swing doors made of wood between the stairs and each storey. They were usually *hooked back to the walls in a fully open position*. The staircase walls were covered with plywood, as were the soffits of the stairs, which also had mineral wool insulation. In the middle of the staircase was a glazed shaft housing the lift. There was an emergency staircase, but of *wooden construction* in a brick shaft with wooden doors. In the original plan of the hotel the corridors in the first to fourth storeys ran right round the building, but at some stage rooms were built in another wing and access to them was through a corridor from an existing wing and adjoining these new rooms were rooms for linen and other stores. The keys to the connecting doors were, it is said, 'available in case of emergency'. A pair of swing doors fitted in the west wing corridors, one pair at each level, *were usually left open.* The corridor walls were covered with 7 mm plywood and entry into the guest rooms was through two sets of doors made of wood. There appear to have been no fire doors, despite the fact that several rooms had recently been modernised.

On the night of the fire nothing at all was noticed until 0225, when a woman guest noticed that the lift was rather warm and close, but took no action. Ten minutes later on the way down again she detected the smell of smoke on the main staircase. She contacted a member of staff and *using the lift* they went to the attic. When they arrived they saw flames that involved the staircase. The guests in the hotel commenced to be warned and the fire brigade was called. Very shortly after this there was a flash over and the fire really took hold. Initially the fire produced a lot of smoke which spread rapidly by the corridors and the main staircase mainly *owing to the constantly open doors.* When the flash over occurred all the combustible materials in the corridors of the topmost storeys and in the main staircase ignited instantaneously. This resulted in a fast, fierce fire centred on the main staircase, with all internal escape routes in the third, fourth and attic storeys blocked by fire.

A total of thirty-five people died in this fire, eleven on the second floor, nine on the third, eleven on the fourth and four on the fifth. All but one of the victims were found to have died from carbon monoxide poisoning. One feature of the fire was the number of persons who died in their rooms even though those rooms were undamaged by the fire. It is thought that the flash over may have built up pressure in the corridors which forced carbon monoxide into the rooms with the result that many of the victims died in their sleep.

Whilst a construction such as described here should not be possible where there are stringent fire regulations, the story illustrates the importance of basic measures such as keeping doors shut and practising a good effective fire drill. For lifts to be used when fire is suspected and more or less certain is inexcusable.

The object of both these illustrations is to bring home the importance of fire prevention, and to show that the measures to be discussed in this chapter are not just an example of government interference, but absolutely necessary measures designed to prevent tragedies such as those described.

Fire precautions

In the UK the Fire Precautions Act 1971 was passed as a means of improving fire safety. At first the Act did not apply to all hotels, but it was amended by the Health and Safety at Work Act 1974, and then by an Order from the Secretary of State in 1976. Fire certificates are now required by all premises where more than twenty persons are employed on the premises at any one time, or more than ten employed elsewhere than on the ground floor. Even if premises do not require a certificate because they do not fit the above, they must supply adequate fire fighting facilities and maintain them ready for use, and ensure that doorways to the outside are unobstructed and easily opened.

In addition to this the 1974 Act also required all premises that came within the auspices of the Offices, Shops and Railway Premises Act 1963 to be equipped with such fire escape facilities for persons employed there as may reasonably be required. However, the important point for the readers of this book is that the obligations for fire precautions are now to be found in the Fire Precautions Act 1971.

Fire certificates. Once a fire authority is satisfied that the means of escape and the related fire precautions are such as may reasonably be required, it must issue a fire certificate. This certificate will specify:

1 the particular use or uses of the premises which it will cover
2 the means of escape in case of fire
3 the method of ensuring that the means of escape can be safely and effectively used at all material times
4 the means for fighting fire for use by persons in the building
5 the means for giving warning in case of fire

The use of the premises as far as a hotel is concerned will be straightforward. The means of escape will, however, be extremely important and this clause will specify the protected routes and their construction. It is not considered necessary to enter into all the various details of this in this book, but rather to stress the duties of management and security officers in ensuring that the escape routes are kept in good order, and that the various measures built into the hotel routine to ensure safe escape are complied with.

Staff training

Managers and security officers will find that their first

most difficult task is the training of staff. One problem, particularly in the large hotel today, is that of language. It is difficult enough to get the various points across, but if the member of staff concerned does not understand English the problem is even worse.

On the initial training day of new staff at least two hours should be spent explaining the fire escape procedures. Staff should be shown the various pieces of fire fighting equipment and be walked round the various escape routes. Then regular training should follow. The recommended level is oral instruction by a competent person for at least half an hour not less then once in every period of six months for employees engaged only on daytime employment and every three months for employees engaged on night duties. To ensure that this instruction is given proper staff records should be kept; these should be kept up to date and be examined by a senior member of management at frequent intervals. It is good practice to stop new members of staff occasionally and pose certain questions such as 'Where is the nearest escape route?' and 'Where is the nearest piece of fire fighting equipment?' Often it is useful to follow this up and ask them how that particular piece of equipment should be used. The answers or lack of them will give a good guide to the level of instruction being given.

Fire drill. In addition to this training, exercises should be carried out once in every six months, including, in larger premises, a simulated evacuation drill with the assumption that one escape route is not available. Again this excercise should be noted on personnel records, and it is better to maintain a record of this fire instruction separately. This record should include the following:

  1 the date of the instruction or exercise

2 the duration of the instruction or exercise
3 the name of the person giving the instruction
4 the names of the persons receiving the instruction
5 the nature of the instruction, training or exercise

Fire doors

It is sad that even in hotels where good training is given, staff still ignore simple notices such as 'Fire door, keep closed'. The example of the fire in Denmark shows that closed doors prevent the spread of fire and this is their purpose. Yet time and time again patrolling security officers will find these doors propped open, often with the nearest fire extinguisher! Leaving doors open is more prevalent in the summer when the building gets warm and staff working on the floors want a good circulation of air. At one London hotel there was a continual battle between security and staff over fire doors leading to a metal fire escape. In summer the patrolling officers would find every door leading to this metal stairway open and tied back with towels, coat hangers or anything else handy. They would remove these and shut the door. By the time they returned to that floor the door would be tied open again. Very rarely did they catch the culprit.

Unfortunately it sometimes takes a fire to bring staff to their senses. In one hotel a small fire started in the basement. It was in some paper and made a lot of smoke. Thanks to open doors, etc., this smoke was drawn up to the seventh floor, where it emerged and set off the smoke detectors. These in turn activated the fire alarm and caused considerable panic amongst guests on that floor, who emerged to find the area full of smoke. They were not to know that the fire was a small one, seven floors away. After this episode doors were shut.

Escape routes

Earlier in the chapter the factors necessary for a fire certificate were set out, the main one being the means of escape in case of fire. In the foregoing paragraphs the training of the staff in escape procedures has been outlined, but equally important is to ensure that the means of escape exist and that they can always be safely and effectively used. It is the latter point that will need the active involvement of managements and security officers.

Escape routes will be nominated in the certificate and will need to be clearly indicated in the building. They will also need to have an emergency lighting system, as a fire may well cause the normal lighting to be extinguished. The doors marked 'Fire door, keep closed', to which we have already referred, will also need to be maintained. These doors may be shut at all times, or they may be a type held back by magnetic catches which are automatically released if the fire warning system is activated.

Protection of escape routes. All escape routes should be protected; that is, they must be separated from the remainder of the building by fire resisting doors (except doors to lavatories) and by walls, partitions and floors of fire resisting construction. In the UK the fire resisting construction must be a type which is tested in accordance with British Standard 476 Part 1 1953 and would have a fire resistance of not less than half an hour. A fire resisting door is defined as a door or pair of doors which, if tested as above, would be free from collapse for not less than half an hour when fitted to its frame and be resistant to the passage of flame for not less than twenty minutes.

Where the fire certificate has been issued these requirements will have been fulfilled, but it is as well for the hotel manager to be aware of the requirements, as often repair and replacement work is carried out in hotels, which may even involve replacing doors to fit in with a new decor. In these cases careful attention must be paid to areas designated for escape, as often the fire resistant requirements can be overlooked.

Outside exits. Patrolling security officers should make frequent examinations of escape routes, paying particular attention to the final escape doors into the street. In these normally unused areas staff will often store unwanted stock, old cardboard boxes etc. Before too long if this practice is allowed to go unchecked the escape doors become blocked or at the very least obstructed, and this type of carelessness can result in injury if a fire breaks out. Checking should not be confined to the inside of the doors as old stock etc. is sometimes stored outside the doors, and if the door is in an alcove property will often be placed back against the door to be out of the weather.

Emergency lighting

Some form of emergency lighting must be provided in the premises and whatever type is used must be capable of illuminating all stairways, exit routes, exit and directional signs sufficiently well to enable people to make their way out of the building. This lighting should by preference be electric. In small buildings, defined as having not more than one floor above ground level and not more than ten bedrooms, electric hand lamps may be accepted.

In larger premises, however, the system should be a permanent installation supplied with electricity from a source independent of the main supply and arranged either to come into operation automatically on the failure of the main supply or to operate at all times when natural light is insufficient for escape purposes. These systems should be capable of maintaining the necessary level of illumination for a period of three hours from the time of failure or disconnection of the normal supply.

The maintenance of this emergency lighting equipment will again be a priority; in the larger hotel it will be in the hands of the chief engineer. Where there is no engineering staff it should be a designated duty of an individual to make periodic checks of this equipment, and this could also be part of a contract with the supplying contractor.

Fire fighting equipment

A further requirement of the fire certificate is that the premises are provided with the means of fighting fire for use by the persons in the premises. The following types of fire fighting equipment will be found in hotels:

hose reels
water extinguishers
foam extinguishers
CO_2 extinguishers
dry powder extinguishers
fire blankets

All such equipment must be regularly maintained by a service contract with the equipment installers. Maintenance is usually carried out annually, and each piece of equipment should carry a label which sets out the dates of the services. These labels should be checked by security or management to ensure that these checks have been properly carried out. In a large hotel with a number of portable extinguishers it is quite possible for some, which may be in out of the way places, to be overlooked. Ill luck is bound to ensure that these extinguishers that have been overlooked will be the very ones seized for use at the outbreak of fire, and will not operate.

All hotel staff should be trained in the use of both the portable equipment and hoses which, in large premises, will be found on each floor. Hoses are mounted on a swivel drum and are simple to operate. However, they usually need to be turned on at the wall point before being taken to a fire and on many occasions hotel staff have been known to rush to a fire with a hose, only to rush back again to turn on the water. It must always be stressed to staff that *under no circumstances should a hose be used on an electrical fire.* If possible the instruction given should be practical, as there is no substitute for actually handling a piece of equipment. There are many makes on the market, which do not all operate in the same way, so the first part of training must concern the actual operation of the equipment. In the large hotel this will entail demonstration because of the language difficulty.

The next important part of the training is the identification of the various types, which should correspond to a colour code: water is red, CO_2 is black, dry powder blue and foam cream. A word of warning is necessary here. Some of the older hotels still have some fire fighting equipment that does not conform to this code; for example, some CO_2 extinguishers were coloured red. In view of the dangers that can arise with colours it is recommended that the hotel equipment be surveyed and that, if any extinguishers are found that do not comply with the colour coding, they should be replaced. This is essential as the instruction must specify which type of extinguisher should be used on which fire. The most dangerous mistake is to attempt to fight an electrical fire with water. Such fires should be attacked with CO_2 or dry powder.

A further point often omitted in such instruction is that the life of the usual type of two-gallon extinguisher is very limited, probably one and a half to two minutes at the most. This is not long when fighting a fire and can create dangers for the operator.

The kitchen staff should receive special attention. The author knows of one case where six ten pound CO_2 extinguishers were used up trying to extinguish a fire that required foam. When a foam extinguisher was used, the fire, which was in a deep fryer, was immediately put out. Such staff should also receive instruction in the use of fire blankets, which must be always held in front of the user as a shield and then dropped over the fire, not thrown up and down as though laying a table! All the latter system achieves is severe burns to the user.

The final point of such instruction is to emphasise that if any portable extinguisher is used it must not be replaced on its mounting hooks, but returned to stores or security, so that it can be replaced. Otherwise when another fire occurs, not much fire fighting can be done with the empty equipment.

All staff working in the hotel should know three things in connection with extinguishers. One, where the nearest piece of equipment is; two, what type it is; and three, how to use it. Management should regularly ask staff these three questions. If a member of staff is unable to answer, then trouble and time must be taken there and then to instruct them.

The final element covered by the certificate is the means for giving warning in the case of fire. Fire alarms must be installed which should be readily available at all times, capable of operation without exposing anyone to undue risk, located throughout the premises, capable of waking staff and residents, and finally, distinguishable from all other signals on the premises.

Where the building has only one or two floors, manually operated sounders may be used provided that any one sounder is audible throughout the premises and that at least two sounders are provided in such positions that a fire occurring anywhere in the premises will not render all the sounders inaccessible through heat or smoke.

Larger premises will have an electrical system, which should have break glass release points. There will also usually be smoke or heat detectors throughout the building which will detect smoke or fire and then raise the alarm.

It is imperative that any such warning system should be in perfect working order as there is nothing worse than continual false alarms, especially if the warning is also given to the fire brigade. Not only does this upset the brigade, but it does nothing for the reputation of the hotel to have firemen in uniform galloping into the premises once or twice a week. The break glass points should be tested weekly, and such tests must be posted in the hotel foyer to warn guests. This testing of the break glass points will also activate the alarm, and will shut doors on magnetic catches where they exist. Any false alarm must be thoroughly investigated.

If smoke or heat detectors develop faults they must be immediately replaced. Hotel management and security officers must know how to do this as it may be necessary to wait for some time for a contractor to attend.

Action in the event of fire

All staff must be aware of the position of any smoke or heat detectors in their area, so that if a fire alarm is made they can check the detectors in their area to see if they have been activated. Normally there is a small red light on the side which will come on if the detector is activated. The fire alarm system itself will also be wired to a central point, and in the larger premises will be zoned. Zoning will help those investigating in that only one area will need to be searched. There should be very clear instructions as to who takes the decisions, and members of senior staff should be given designated tasks to follow should a warning of fire be received. Certain members of staff will be designated as fire fighters and others assigned to warn guests and assist with evacuation. Not to be overlooked is the task of calling all the lifts to the ground floor and ensuring that they stay there, a job normally undertaken by the hotel porters.

Any fire in a hotel, however small, should be treated very seriously, and no fire should be accepted as having started accidentally until a full investigation has been made in conjunction with the fire brigade. A very careful record should be kept of all fires, listing not only the time, date and type of fire, but the person reporting the fire, the person who extinguished it and others who were at the scene. It may seem tedious to record these details, but this type of information will often point to an arsonist. In one recorded case small fires were found in rubbish sacks on floor landings. The first one was accepted as accidental and believed to have been started by a cigarette end from an ash tray, but then a further one occurred, followed by two together. Fortunately these fires were on concrete landings and could not have caused damage to the hotel itself, but nevertheless smoke was produced, the alarm system was activated and some panic resulted each time. A perusal of the records showed that one member of staff seemed to be present at each fire. He was subsequently interviewed by police and charged with arson, and pleaded guilty in court. The danger of arson must never be overlooked.

With the rubbish sack fires the hotel was fortunate, but small fires could easily have been started in a linen cupboard or other area, and then the consequences could have been horrific.

Fire is an ever-present danger, which each year causes far more loss than theft, and many, many deaths. Hotels in the UK have a good record compared to those in many other countries. This is in no small way a result of the excellent legislation, but the law will only be effective if staff are aware of the problems and treat the various regulations seriously, not just as a nuisance.

12 Terrorism and disasters

It is a sad reflection of our society today that it is necessary even to include this chapter. A few years ago the prospect of terrorists attacking a hotel by bomb or other means would have been unthinkable, especially in the UK. The situation in Northern Ireland has changed that and the terrorist activity that unhappy country has suffered has crossed the Irish Sea. In addition, terrorist methods have been adopted by numerous groups all over the world, and any premises or people might find themselves under attack for reasons they do not even understand. Bombs and bullets do not differentiate between the intended victim and the innocent.

Bombs and other terrorist attacks

The terrorist's most common attack is by bomb, but the use of firearms is not unusual, especially in attacks by determined groups such as that on the Iranian Embassy in London in 1980. The bomb attack, however, remains the most likely danger and the one that should be uppermost in the mind of hotel managements when considering this subject. A bomb constitutes not only a danger to the lives of staff and visitors, but danger to the property and a danger to the hotel's continued ability trade and thus the livelihood of its employees. The threats of such attacks are also designed to disturb, frighten and interfere in the economy, and such threats are quite commonplace today.

Just as fires will occur when least expected, so also will the bomb attack or the bomb threat. In the same way as hotel management are trained in security (partly by this book!), so does the terrorist receive training, and this training will teach him to strike when his victim least expects it and is therefore most vulnerable. If one studies the pattern of bomb attacks in London for example, one sees an outbreak then a long interval, followed by another outbreak when the terrorist feels that everyone's guard is again relaxed. The time to strike a hotel is when it is very crowded or, conversely, at the quiet times when staff levels are low.

Bomb threat calls

The majority of bomb threats are made by telephone, as are all genuine warnings that a bomb has been planted. Let us hope that we do not reach a situation where no warning is given, but that possibility should not be overlooked. For the moment, however, concentration must be focussed on the bomb call that is received and the action that should follow. In most hotels the call will be taken by a girl on the hotel switchboard, and the action that she takes or does not take will have great importance. Hotels that receive a number of these calls develop switchboard staff with an expertise for sorting out the false calls, but one cannot rely on this and therefore proper training must be given to those staff likely to be on the receiving end of such calls.

The first aspect of the training to be stressed is 'Don't panic'. Calmness is essential. Staff should be instructed to try to keep the person talking as long as possible in order to gain all possible information, not only about the alleged bomb but also about the caller. They should ask 'Where is it?' 'Why have you done this?' The chances are that the caller will ring off as soon as the threat has been made, but this is not always the case, particularly with a hoax call. The switchboard operator should inform the caller that the hotel is occupied and that his action could result in death or serious injury. There may or may not be a reaction to this, but it is always worth a try if time is given. The operator should pay particular attention to any background noises such as motors running, background music, airport noise, train noise etc., indeed any sound which may give a clue to the location of the caller. They should also listen carefully and closely to the voice and judge its quality, the sex of the caller, and local accents, dialects or speech defects. The above methods will soon show whether it is a hoax call by children, but the operator must never assume that a call is a hoax. All calls must be treated seriously.

As an aid to operators it is useful to have a printed form available for such cases, setting out all the points referred to above, which they can use to note the various points as they put the questions to the caller. A suitable form for this purpose is shown on pages 124–25.

Where there is more than one switchboard operator there should be a pre-arranged signal to alert all operators that such a call is being received. This will enable any agreed plan to be put into operation and management to be informed at the earliest opportunity. Every second counts in such calls, for if the call is genuine and there really is a bomb on the premises then every available minute will be needed.

There must be a nominated member of staff to deal with all such calls that are received, the chief security officer if there is one on duty, otherwise the duty manager. It must, however, always be one person as it is not a responsibility to be shared out. A decision must be made by one person. The first thing the person responsible must decide is how seriously to take the call, a decision which will be helped considerably by the manner in which the call has been taken, and the background information that the operator can give him. Also he will, of course, be aware of the current situation; for example, if there is a spate of bombings at the time then obviously the call will be taken more seriously than if there have been no recent attacks.

Evacuation or search

If the nominated person decides to take the call seriously, then his next decision must be what action to take. This decision will be guided to a certain extent by the supposed location of the alleged bomb and what time it is due to go off. There may be other relevant circumstances to consider. If the alleged bomb is due to explode very soon, then he may have no alternative but to order a complete evacuation. If the explosion is not due for some time he may decide to mount a search operation, or he may carry out a combination of the two. The following case may well serve as a guide to the way the decision maker should think.

The hotel concerned was in London and at the time was privileged to have as guests two Prime Ministers atttending the Commonwealth Prime Ministers' Conference. This is obviously an unusual occurrence and such circumstances must affect decisions. At about 5 pm a call was received by the hotel switchboard stating that two bombs had been placed on the premises and would explode at 5.45 pm. The girl taking the call was an experienced operator who felt that the call could be genuine. She passed the call to the Chief Security Officer and the Duty Manager, who were walking the floors together at the time. They were just debating the situation when a pageboy came walking along the corridor on the floor they were patrolling. He had with him two plain brown paper parcels. They stopped him and asked him what they were, but he did not know. He said he was taking them to the suites occupied by the Prime Ministers. The parcels were taken from him and carefully placed in a store room on that floor without windows and with the door locked. The parcels had apparently been received at the hall porters' desk, but nobody there could say who had brought them in except that they were of Indian appearance and had just said the parcels were for the two Prime Ministers. These two facts were sufficient for immediate action to be taken. The floors above and below the store room and that floor itself were evacuated, police were informed and the bomb squad attended. The story ended happily when the deadline passed and the parcels were opened by the bomb squad and found to contain a present of mangoes. However, the circumstances surrounding this are a clear illustration of how such events must be taken into account when action is taken. Obviously, in this case, the call and the parcels were a strange coincidence but there is no doubt that the appropriate course was followed.

To enable correct action to be taken quickly, certain staff should be nominated as a search party for each floor. They should then get to know every aspect of their area so that they will see anything out of the ordinary or suspicious. If a decision is taken to search then these staff should commence their patrol, reporting immediately if anything suspicious is found, and at the end of the search if they find nothing.

Police should be informed at the earliest opportunity, but they will not necessarily attend. Many bomb threat calls are received and, if police attended every call, they would do nothing else. If the situation develops into a real threat, however, they will personally attend and will ensure that the bomb squad are informed.

At some stage management will be faced with the decision whether or not to evacuate. If a bomb or a suspicious package is found then the decision is simple, but often it will need to be made without such evidence. No real guidelines can be laid down for this. Each case, like the one outlined above, has to be decided on its merits. If a decision to evacuate is made, then a set plan should be put into operation to enable evacuation as quickly as possible and with the minimum of fuss. It is debatable whether guests should be told the reason for the evacuation, as panic must be avoided at all costs. It is far better to use the fire alarm as an evacuation warning. For this reason we should reiterate what was stated in chapter 11, that false fire alarms of any frequency are a problem. Where many false alarms do occur it is very difficult to convince people that an alarm is genuine.

The evacuation should be directed into designated areas at a distance from the hotel and away from places likely to cause injury such as plate glass windows. One problem that will be encountered will be the difficult person who will refuse to evacuate, often a member of staff. No refusal can be accepted and this is one occasion when extremely firm handling is required.

All the above procedures should be clearly set out in a manual available to all senior staff, and each member of staff should be aware of his or her responsibilities and should receive instruction accordingly. The manual should be kept up to date by the chief security officer or by the senior manager, if there is no security. It must be issued to all new senior staff, and as each nominated person leaves a new nominee must be appointed. The evacuation procedure must be practised, possibly in conjunction with fire training. An important part of the practice will be to have sufficient staff on duty to direct guests to the various escape routes, and to remain on each floor until it is clear. It is never possible in a hotel to know whether guests are in residence or not, so entering and checking all rooms must be part of the evacuation procedure.

Letter bombs

Another type of bomb attack that can be expected is the letter or parcel bomb. This is a favourite system throughout the world; in 1972, for example, letter bombs were received in Brussels, Buenos Aires, Geneva, Jerusalem, London, Montreal, New York, Ottawa, Paris, Phnom Penh, Tel Aviv, Kinshasa, Beirut, Tripoli and Algiers. The increase in this type of attack using the postal service to deliver the bomb poses many serious security problems. Hotels are not immune to letter bombs and must ensure that their bomb threat plans include the threat of letter bombs.

It will be normal practice for one person or department to handle the incoming mail, and the relevant staff must be made aware of the dangers. Posters can be obtained for the wall of the mail room or office used outlining the various points to look for when sorting the mail.

Letter bombs have usually taken the form of substantial envelopes or parcels containing paperback books. They will probably be in the form of a flat letter and will weigh up to 4 ounces, or can be a parcel about the size of a paperback book. There are several points which should make the sorter suspicious. These are:

1 the postmark – if unfamiliar or foreign
2 the writing – lack of literacy – crudely done etc
3 personal or private letter addressed to senior managers under title, not name
4 incorrect spelling of the name and address
5 excessive weight for the size
6 grease marks on the wrapping which seem to come from inside
7 smell – some explosives have a smell of almonds or marzipan
8 an abnormal fastening

9 any sign of wires or batteries

The instructions to the staff should make quite clear that in no circumstances should they attempt to open any such parcel or package. They should isolate it where it will do the least harm, keep people away from it and open all doors and windows in the vicinity. Under no circumstances should they try to make such a parcel safe by putting it in water, which could prove very dangerous. Police should be informed at once and then everyone should keep away from the object until police arrive.

Radio-controlled bombs

One point needs to be stressed with all hotel staff when dealing with any type of bomb threat or bomb. Many of the now very sophisticated bombs in use are detonated by remote control, the bomb that killed Airey Neave at the House of Commons being a case in point. This type of bomb can be accidentally activated by another radio nearby, or by the 'bleeper' type of device carried by senior hotel staff. Should any bomb threat be received all remote-controlled devices in the area should be switched off or taken away.

Hoaxes

The vast majority of bomb threat calls received and some of the devices in the post are hoaxes. Hotel management should be aware that since 1977 this type of hoax is a specific offence in the UK. The Criminal Law Act 1977 section 51 says:

Any person who places any article in any place whatsoever or dispatches any article by post, rail or other means with the intention of inducing in some other person a belief that it is likely to explode or ignite and thereby cause personal injury or damage, is liable to a penalty on summary conviction of imprisonment up to three months and or a fine of £1000.

Firearm attacks and kidnappings

Terrorist activity may of course, not be confined to bombs and we have already discussed the possibility that firearm attacks may be made. There is very little the hotel manager can do to guard against this type of attack except to be aware of the problem, and to keep a careful eye on his guest list as this may often be a guide to a possible attack. Another attack is kidnap with a view to extorting money or securing the release of a comrade. Hotel staff are unlikely to be targets, but the chief executive of a hotel chain might well be, particularly where the hotel chain is an international one. The hotel may well have a guest who is liable to such attack. Again there is not a great deal the hotel can do, but they can try to minimise the threat by asking that such persons do not book into the hotel in their own name but that any booking be done in the name of the company involved. Terrorists often rely upon innocent information received via hotel staff, and a well known personality will be the subject of comment by staff, which can be kept to a minimum if the booking is not made in his name. Visitors demanding to see such a person should be carefully monitored and not allowed up to the room before a check has been made. Any knowledge of movements in or out of the hotel should be kept to as few

people as possible and should not be advertised in any
way. If police are not already aware of the visit they
should be informed.

Disasters

Here we mean a very serious fire with heavy loss of life,
an explosion resulting in great damage and injury,
flooding, an air crash or other incident which involves
loss of life. Disasters can happen at any time and can be
man made or an act of God such as an earthquake or
hurricane.

In the event of a disaster the emergency services of
police, fire and ambulance will very quickly be in
attendance and the overall running of the incident will
probably be in the hands of the police. However, in the
first few minutes of a disaster certain action needs to be
taken well before the emergency forces have arrived, and
this preliminary action should be foreseen and a plan be
made for action if and when required. In a hotel the
preliminary action will centre around evacuation, but
should also cover such essential information as will be
required by the emergency services, such as the position
of the mains supply of gas and electricity. All senior staff
must be familiar with these details, for if the information
is only held by for example the engineer, it will be of no
use, should he be buried under the nose of an aircraft
that has crashed on part of the hotel. The depth of this
planning will not be as great as for example on a large
industrial site, but it should be considered and any plans
felt necessary should be drawn up.

The whole point about any disaster, bomb threat,
terrorist attack or similar event is that the danger must
be accepted and the possibility considered. No person or

business is immune from such attacks today, and however remote the possibility management would be failing in its duty if it did not plan for such emergencies. It is a very important aspect of security and all procedures should be regularly reviewed and brought up to date. Often excellent procedures are laid down but then forgotten, and when they do have to be put into operation they are found to be out of date and in many cases inapplicable. If loss of life resulted from this management would, quite properly, suffer adverse criticism.

Form for recording details of a bomb threat

Date............................ Exact time ....

Exact words of caller ....

....

....

Was any person requested by caller by name? ....

If so who? ....

When will it explode? ....

Where have you placed it? ....

What does it look like? ....

Why have you done this? ....

Who is calling? ....

Whom do you represent? ....

Time call terminated ....

The caller

Sex .... Approximate age ....

Local/long distance call ....

Voice: Fast .... Slow .... Stutter .... Distinct ......

Disguised .... Educated .... Foreign accent ...

Loud .... Soft .... Local accent ...... Dialect ....

Manner: Calm ... Angry .... Emotional .. Laughing ...

Deliberate ....

I can/cannot imitate characteristics of the caller's voice.

The voice was/was not familiar to me.

Background noises ....

Management notified .... Time ....

Police notified .... Time ....

Signed .... Date ....

Time ....

13 Training in security

There is a considerable body of opinion even today amongst senior managements that security can be carried out by almost anybody. Often the security function is allocated to one man called the chief security officer or the security manager, while the rest of the staff sit back and leave it all to him. Hotels are not immune to this thinking and often forget about security once the chief security officer has been appointed.

The chief security officer

As we have seen from earlier chapters the appointment of the security officer is of prime importance and it is an appointment that requires very careful selection. A considerable number of hotel chief security officers are retired police officers, and it is obvious that hotel management believe that the appointment of such a person will solve their problems and nothing more will be required of them. No thought is given to the training of the man or his staff. Hotels are fortunate in that most of the men so appointed have been hard working, conscientious individuals who have taken the trouble to learn their new trade – and it is certainly a new trade for the retired police officer. He is leaving a public service for a private one, entering the commercial world for the first time with all that that implies, and dealing with facets of security he will not have touched on in the police service unless he has served as a crime prevention officer.

It should always be a rule of a hotel that any security officer appointed to the job for the first time will receive training in that task. That training can be given in a number of ways and will probably need to be split between the hotel and outside bodies. It is doubtful whether hotel training officers have the expertise to deal with security and law, but they can handle an introduction to hotel life, the hotel systems, the documentation in use, any hotel disciplinary codes and the hotel management and departmental structure. Time should also be given to explaining the various terms in use in a hotel. Each trade or profession has its own peculiar jargon and the hotel trade is no exception. The security officer will deal with all departments with all their different problems and will need to know what he is taking on. In chapter 8 two cases of fraud were outlined, and the fact that the various control methods of dealing with documentation were circumvented in these cases shows that knowledge of the systems and how they work must be an integral part of hotel security training. Usually the poor security officer has to find out about the systems as he goes along, and often only learns about them when there is a problem.

The internal hotel training should include attachments to all departments for a short period, both to get to know the department heads and to become familiar with the operations of the departments. It is not suggested, however, that the security officer actually work in the department, but rather be schooled in all its operations. He will find such an exercise of great benefit, and learn far more from it than from any formal instruction in a classroom. All this training should be completed in a period devoted entirely to training, and not fitted in alongside normal duties. With many different demands on his time the new security officer will never complete his training programme.

Security training. The second part of the security officer's training should ensure that he is aware of the security function, and cover those parts of the law which he may not have come across as a police officer but will need to know in his present task. The author is not aware of any course designed to fit this requirement, and it may be that the security officer in the UK will need to attend one of the excellent courses run by the International Professional Security Association. These are run on three levels; basic, intermediate and advanced. The basic course is really designed for the man or woman who has never been employed in either security or police work and usually lasts three days. The intermediate is designed for the more experienced security officer who wishes to progress or who is earmarked for promotion by his company. The advanced course is designed for the chief security officer or security manager and deals with security at that level. This course is residential and of one week's duration.

The subjects that a hotel security officer will need to know about are:

law of trespass
criminal law – Theft Act, criminal damage, assault
   Hotel Proprietors Act, Aliens Order, etc.
powers of arrest
search procedures
laws of evidence, Judges' rules
court procedure
use of pocket book and report writing
bomb threat procedures
patrolling of premises
principles of first aid
elementary fire fighting

The retired police officer should be well abreast of much of this but will still benefit from some revision.

An alternative means of giving the training would be for the hotel to retain a qualified security consultant, well versed in the problems of hotel security, who could undertake such training as and when required, as well as the other training which will be covered later in this chapter. Another good method of arranging the practical side of this training is for hotels to have a mutual aid scheme whereby an inexperienced security officer from one hotel is attached to an experienced officer from another. This can obviously work very well with the hotel chain, and could be a consideration for a number of small hotels who only employ one person for this purpose. As a final point in this section, British hotel management should note that any person who applies to them for a post in security and is qualified as a Graduate, Member or Fellow of the Institute of Industrial Security will have taken the Institute's examinations and will be a well qualified person who will have studied most of the subjects referred to, and will need very little training, if any, in security subjects.

Hardware. The final aspect of any hotel security officer's training should cover the hardware: safes, locks, alarm systems and so on. The officer must have a good knowledge of these and be able to tell when a lock is a good one, and when it is the right lock for the door. He should know about the construction of safes and the types that are available. He should have a basic knowledge of the working of intruder alarm systems and be aware of the current trends in the market. If a safe, lock or alarm is required in a hotel it is the security officer who will be expected to advise on the correct one and without this knowledge he could give the wrong advice and cost the hotel unnecessary money. The person with an Institute of Industrial Security qualification will already have this knowledge, others will need to be taught.

The above paragraphs will have shown that the training of a security officer for a hotel is considerable, but if it is done in the way suggested then the benefits to the hotel will be endless, not just the added interest of the individual himself. It is very disturbing to anyone appointed to such a position to find himself thrown in at the deep end and faced with all sorts of problems, without the benefit of training or, therefore, any idea of how to tackle the problems. No doubt a competent individual will get through, but at what cost to the hotel?

Other hotel staff

Security training does not stop, of course, at the hotel security officer, but should be an important part of the training of all hotel personnel. The training of other hotel personnel can be undertaken by the hotel chief security officer if he is competent, but it should be remembered that not everyone is a natural lecturer, or a willing one, and it is well for this aspect of the job to be considered when the chief security officer is appointed. The level of security training to be given will vary with the task of the individual. The training required by a trainee manager, for example, will be far greater than that of the chambermaid. The suggested levels of training would follow a pattern of:

managers and heads of departments
departmental supervisor level
other staff

Senior staff

The managers and heads of department should receive

security training that will cover a syllabus based on clearly defined subjects. Departmental supervisors would cover the same syllabus but at a lower level of information. The length of the training will vary according to the amount of information covered, but should not be less than one day. In existing hotels where very little training has been given the whole staff could be trained in a series of courses, using outside lecturers.

The subjects to be covered in the lectures to managers should include:

a general outline of crime prevention and the present position of hotel security

an outline of hotel legislation as it affects security, and the various legislation surrounding reservations

the law of trespass and the powers of hotel staff to take action against undesirables such as prostitutes and drunks

the security of rooms, including the hotel system for dealing with room keys, the security of valuables and the handling of found property

the various control systems in operation to supervise the operations of the bars, restaurants and the floor service etc.

a reminder of the fire prevention legislation and a discussion of bomb threats and the various counter measures to be taken

the problem of 'walk outs', how these can be prevented and what action to take

Other subjects can be added to this as considered necessary, but the above should form the nucleus of the course. The depth at which it is studied will vary between managers and supervisors.

It would be an advantage to involve the local police in this type of course. Normally the local crime prevention officer will only be too delighted to come and talk to the courses about his subject and update them on the current crime situation in hotels. In London, the Metropolitan Police Hotel Squad specialise in hotel crime and may be glad of the opportunity to outline to hotel staff the various problems that they encounter. Some practical work should be included in the course: this can be done by setting up a room theft, then seeing how many of the relevant details are obtained by the students. Another exercise could concern a bomb threat telephone call. Plenty of discussion periods should also be brought in as it is from these that personnel's past problems will emerge.

Junior staff

The training of other staff need only concern their own particular department and will probably be covered in a period of some two hours. It is best given by a member of the hotel staff, preferably the chief security officer. It should be borne in mind that many of the staff may have little or no knowledge of English: visual aids will need to be much in evidence. The lecture should concentrate on individual work areas and the problems that each member of staff may encounter. As an example, the housekeeping staff will need to be fully aware of the various security systems in operation with the rooms, will need instruction on the proper securing of rooms whilst they are working on them, and will need to know how to deal with found property. Hotel porters will need to concentrate on spotting undesirables entering the hotel, and the appropriate action.

In all this training great emphasis must be placed on the effect of breaches of security on the hotel's reputation. The staff should be aware that, if the hotel gains a bad reputation, their jobs are at stake; this point can have far more effect than any other measure. No one, especially today, wishes to lose his job and job security should be an incentive to make sure that security procedures are followed. Training is an important part of management and if a positive training policy is carried out the hotel will gain immensely.

Appendix: *The law relating to hotels in the UK*

The law relating to hotels is extensive, and many Acts of Parliament have some bearing on the activities of a hotel. This appendix will concentrate on those aspects of the law which, it is felt, affect security. Those who wish to make a more detailed study of the law as it affects hotels and catering are referred to *Hotel and Catering Law* by David Field, published by Sweet and Maxwell.

Fire

The Fire Precautions Act of 1971 has already been covered in chapter 11, as also have the implications in that respect of the Health and Safety at Work Act 1974.

Obligations of the hotel proprietor

The one Act that specifically applies to hotels is the Hotel Proprietors Act 1956. Section 1 of that Act defines the establishments that are covered and says this:

A hotel within the meaning of this Act shall, and any other establishment shall not, be deemed to be an inn; and the duties, liabilities and rights which immediately before the commencement of this Act by

law attached to an innkeeper as such shall, subject to
the provisions of this Act, attach to the proprietor of
such a hotel and shall not attach to any other person.

The proprietor of a hotel shall, as an innkeeper, be
under the like liability, if any, to make good to any
guest of his any damage to property brought to the
hotel as he would be under to make good the loss
thereof.

In this Act the expression 'hotel' means an
establishment held out by the proprietor as offering
food,   drink   and,   if   so   required,   sleeping
accommodation, without special contract, to any
traveller presenting himself who appears able and
willing to pay a reasonable sum for the services and
facilities provided and who is in a fit state to be
received.

Comment: The term 'held out' is a vital element and
means that all the rights and duties under the Act fall
upon the proprietor, who may, of course, be a company
rather than a person. The duties laid upon the proprietor
are owed by him to the person who can be classed as a
'traveller'. This is important as in the case of R *v* Higgins
in 1948 it was held that an innkeeper may not refuse to
supply a 'traveller' with food and lodging without lawful
excuse. It was held in this case that it is a question of fact
on a full review of all the circumstances whether there
has been an unreasonable refusal to supply a traveller
with food.

The term 'traveller' does not appear to be restrictive
in its meaning and can apply to anyone, irrespective of
his distance of travel or his length of stay. However, in
the case of Lamond *v* Richards it was held that a
'traveller' ceases to be such after a reasonable time
(having regard to the circumstances) and may be
required to leave an inn after reasonable notice.

In an old case, R *v* Ivens 1835, but one which is still of

interest it was said: 'An Innkeeper is not to select his guests. He has no right to say to one "You shall not come into my Inn" and to another "You shall", as everyone coming, *and conducting himself in a proper manner* has a right to be received.' In 1951, in the case of Williams v Linnit it was decided that a man calling in at a hotel for refreshment on his way home from work was, in the eyes of the law, a traveller.

The first of the proprietor's duties is to provide food and drink, but only such food and drink that he has available and what may be considered reasonable in the circumstances. He would be quite right to give priority to his existing guests, and to refuse to supply the traveller if the regular guest would be deprived as a result. His second duty is to supply accommodation, but obviously he can only supply what is available. If the hotel is full he can refuse to supply accommodation and under these circumstances the person demanding cannot, for example, demand that a camp bed be erected in an office.

Section 1 also allows the proprietor certain other grounds for refusing. These are that the person is not willing and able to pay a reasonable sum for the services and facilities provided, or that he or she is not in a fit state to be received. A simple example of the latter would be a drunk. There is nothing in the Act to prevent the proprietor from demanding payment in advance, and the Act has no control over room allocation.

Discrimination

The proprietor's right to refuse to supply services on certain grounds can be affected by two other Acts of Parliament.

Sex Discrimination Act 1975

It is unlawful for any person concerned with the provision of goods, facilities or *services* to the public or a section of the public, to discriminate against a woman who seeks to obtain or use these goods, facilities or *services* by refusing or deliberately omitting to provide them or by refusing or deliberately omitting to provide her with goods, facilities or *services* of the like kind and on the like terms as are normal in the case in relation to other members of the public. (section 29)

Race Relations Act 1976

It is unlawful for any person concerned with the provision of goods, facilities or *services* to the public or a section of the public to discriminate against a person who seeks to obtain or to use those *services* by refusing or deliberately omitting to provide him with any of them or by refusing or deliberately omitting to provide him with goods, facilities or *services* of the like quality, in the like manner and on the like terms as are normal in the first mentioned person's case in relation to other members of the public. (section 20)

Comment: It will be noted that there is very little difference in the wording of both Acts; the important point is discrimination. To discriminate is defined in the Race Relations Act as:

To treat a person less favourably than others on the grounds of colour, nationality, race or ethnic or national origins, or imposing conditions which such a person is less likely to fulfil than others and which cannot be justified on non-racial grounds.

Loss of or damage to guests' property

Of extreme importance for security – the proprietor's liabilities in this respect are clearly set out in section 2 of the Hotel Proprietors Act 1956.

Without prejudice to any other liability incurred by him with respect to any property brought to the hotel, the proprietor of a hotel shall not be liable as an innkeeper to make good to any traveller any loss of or damage to such property except where –

(a) at the time of the loss or damage sleeping accommodation at the hotel had been engaged for the traveller; and

(b) the loss or damage occurred during the period commencing with the midnight immediately preceding, and ending with the midnight immediately following, a period for which the traveller was a guest at the hotel and entitled to use the accommodation so engaged.

Without prejudice to any other liability or right of him with respect thereto, the proprietor of a hotel shall not as an innkeeper be liable to make good to any guest of his any loss of or damage to, or have any lien on, any vehicle, or any property left therein, or any horse or other live animal or its harness or other equipment.

Where the proprietor of a hotel is liable as an innkeeper to make good the loss of or any damage to property brought to the hotel, his liability to any one guest shall not exceed fifty pounds in respect of any one article, or one hundred pounds in the aggregate, except where –

(a) the property was stolen, lost or damaged through the default, neglect or wilful act of the proprietor or some servant of his; or

(b) the property was deposited by or on behalf of the guest expressly for safe custody with the proprietor or some servant of his authorised, or appearing to be authorised, for the purpose, and, if so required by the proprietor of that servant, in a container fastened or sealed by the depositor; or (c) at a time after the guest had arrived at the hotel, either the property in question was offered for deposit as aforesaid and the proprietor or his servant refused to receive it, or the guest or some other guest acting on his behalf wished so to offer the property in question but, through the default of the proprietor or a servant of his, was unable to do so.

Provided that the proprietor shall not be entitled to the protection of this subsection unless, at the time when the property in question was brought to the hotel, a copy of the notice set out in the Schedule to this Act printed in plain type was conspicuously displayed in a place where it could conveniently be read by his guests or near the reception office or desk, at or near the main entrance to the hotel.

Comment: The above quite clearly sets out the liability of the proprietor in respect of loss or damage to guests' property. He must ensure that (a) there is no neglect by his staff, and that (b) the statutory notice is prominently displayed. This should read:

NOTICE
Loss of or damage to guests' property

Under the Hotel Proprietors Act 1956, a hotel proprietor may in certain circumstances be liable to make good any loss or any damage to a guest's property even though it was not due to any fault of the proprietor or staff of the hotel.

This liability however
(a) extends only to the property of guests who have engaged sleeping accommodation at the hotel;

(b) is limited to £50 for any one article and a total of £100 in the case of any one guest except in the case of property which has been deposited, or offered for deposit, for safe custody;

(c) does not cover motor cars or other vehicles of any kind or any property left in them, or horses or other live animals.

This notice does not constitute an admission either that the Act applies to this hotel or that liability thereunder attaches to the proprietor of this hotel in any particular case.

Safety of the guest

This has always been recognised even at common law and many cases in the past stressed this fact. In Sandys *v* Florence in 1878 it was said 'Whilst the guest is in the house the innkeeper is bound to take reasonable care so as to prevent any danger to him'. However, in the case of Walker *v* Midland Rail Co 1887 it was held that the duty does not extend to every room in the house, at all times of night and day, irrespective of whether any such guests may have a right, or some reasonable cause, to be there; but it is limited to those places into which guests may reasonably be supposed likely to go, in the belief reasonably entertained that they are entitled or invited to do so.

The above has been emphasised by the Occupiers Liability Act 1957. Section 2 of this Act makes the occupier of premises liable for the physical safety of *all persons lawfully entering the premises* and in addition he is under a duty of care to ensure that the premises are fit for the purpose for which the visitors are invited to use them.

Registration

The registration of guests at hotels is now dealt with by the Immigration (Hotel Records) Order 1972 which reiterated Article 19 of the Aliens Order 1953. This Order says:

> Every person of, or over, the age of 16 years, who stays for one night or more at premises (furnished or unfurnished) where lodging or sleeping accommodation is provided for reward, must furnish on arrival his full name and nationality to the keeper of the premises.
>
> Aliens must, in addition to the above, provide the number and place of issue of their passports, certificates of registration or any other document which establishes identity. Before leaving the keeper must be informed of the address of their next destination if known.
>
> The keeper of such premises must maintain a record of this information and keep it for at least 12 months. This record must be made available for inspection by a constable or other authorised person at all times.

Comment: The important point of the above is that this Order applies to *all persons,* whether British subjects or not. Such records must be kept for at least twelve months.

Trade descriptions

Another aspect of registration of note to security relates to the provisions of the Trade Descriptions Act 1968. Under section 14 of this Act it is a criminal offence knowingly or recklessly to make any statement in the course of any trade or business which falsely describes any services or facilities offered.

The Act itself provides a defence to the above by virtue of Section 24, which says that any person who can show that

(a) it was a mistake on his part; or that
(b) he placed reliance on information supplied to him; or that
(c) the offence was the result of the act or default of some person,

will have a defence. Any complaint, therefore, about the hotel facilities or services should be studied with this Act in mind, particularly where the hotel brochure is concerned.

Right of lien

Although not the subject of an Act of Parliament, the right of lien over property is still a very important right held by an innkeeper, and therefore a hotel proprietor if he comes within the definition. The right of lien is a right to detain property until a bill is paid in full. The right covers all items of property which appear to belong to the guest, and which are of the type one would normally expect the guest to have with him. Two cases of importance should be mentioned. The first is Robins v Gray in 1895. In this case the right of lien was held to extend to sewing machines brought to the hotel by a commercial traveller, on the grounds that he would normally be expected to have with him the product that was marketed by his employer.

The second case was Marsh v Commissioner of Police in 1945. The circumstances of this case are remarkable. A guest stole a ring from a local jeweller during his stay at the hotel, and then found himself unable to meet his hotel bill. He then handed over this stolen ring as security against the non-payment of the bill. He was subsequently convicted of theft and the jeweller sued for the return of the ring. The court held that since the hotel thought in good faith that the ring was the guests' property, they had a right of lien in it.

The Hotel Proprietors Act did, however, cancel the right of lien that used to be held on cars, property left in cars, horses, horses' harness and other animals (section 2).

Comment: Today it is worth bearing in mind that the Theft Act 1968, section 28, says 'Where goods have been stolen, and a person is convicted of an offence with reference to the theft – the court may order anyone having possession or control of the goods to restore them to any person entitled to recover them from him'. It is unlikely therefore that the circumstances of Marsh above will be repeated.

Criminal law

Security will be concerned with any criminal offence that may be committed on the hotel premises, and also offences which may require them to take some action before the police arrive. They will need, therefore, some knowledge of those parts of criminal law which may concern them.

Although at present not a criminal offence in itself, trespass is important, more particularly since the passing of the Theft Act 1968. The simple act of trespass itself is, of course, a civil matter; it is the act of being upon land or property without any right to be there. Any reader wishing to increase his knowledge of the civil law surrounding trespass should refer to *Winfield on Tort,* published by Sweet and Maxwell.

Trespass is now an important part of the offence of burglary, which is defined in section 9 of the Theft Act 1968:

> A person is guilty of burglary if –
> (a) he enters any building or part of a building as a *trespasser* and with intent to commit any such offence as is mentioned in subsection 2; or
> (b) having entered any building or part of a building as a *trespasser* he steals or attempts to steal anything in the building or that part of it or inflicts or attempts to inflict on any person therein any grevious bodily harm.

Subsection 2: The offences referred to above are offences of stealing anything in the building or part of a building in question, of inflicting on any person therein any grievous bodily harm.

It is apparent therefore, that this is an offence that could be committed in a hotel, and it is important in that it is an offence that carries a power of arrest for any person. Section 1 of the Theft Act 1968 defines theft thus:

> A person is guilty of theft if he dishonestly appropriates property belonging to another with the intention of permanently depriving the other of it.

This again is important in that there is, as will be seen later, a power of arrest for the private person and thus the hotel staff.

Obtaining services by deception

Section 1 of the Theft Act 1978 says:

> Any person who by any deception dishonestly obtains

services from another shall be guilty of an offence.

Obtaining of services is defined as:

> An obtaining of services where the other person is induced to confer a benefit by doing some act, or causing or permitting some act to be done, on the understanding that the benefit has been or will be paid for.

Comment: The above covers the 'walk out' type of offence. To prove the offence dishonesty and the practising of a deception are necessary, and, of course, the obtaining of a service.

Also contained in the 1978 Act was provision to deal with the offence of consuming a meal in a restaurant and leaving without paying. This offence has proved difficult to prosecute over the years and is now dealt with by section 3 of the Theft Act 1978:

> Any person who knowing that payment on the spot for any goods supplied or services done is required or expected from him, dishonestly makes off without paying as required and with intent to avoid payment shall be guilty of an offence.

The above is now a very straightforward offence and should not create difficulties. However, as with many criminal offences it is proving the 'intent' that needs the closest attention.

In order to prevent the offender escaping before the arrival of police in such cases, section 3 subsection 4 says the following:

> *Any person* may arrest without warrant any person he suspects with reasonable cause to be committing or attempting to commit an offence under section 3.

Arrest

There has always been a power of arrest for the private

citizen under common law, and this power was brought into the Criminal Law Act of 1967. It is a power that should be known to hotel staff, particularly the security officer, but no arrest should be made unless the staff are certain of what they are doing. The relevant parts of section 2 of the Criminal Law Act 1967 say:

Section 2 (2) Any person may arrest without warrant anyone who is, or whom he with reasonable cause suspects to be in the act of committing an arrestable offence.

Section 2 (3) Where an arrestable offence has been committed any person may arrest without warrant anyone who is or whom he with reasonable cause suspects to be guilty of the offence.

Comment: The first and most important condition of this power is that there has to be what is termed an 'arrestable offence'. For present purposes this can be defined as an offence for which a person not previously convicted can on conviction or indictment (at the crown court) be sent to prison for five years. Thus it is the term of imprisonment that can be given that is important. If we refer to the criminal offences set out above, burglary and theft are both arrestable offences, as is obtaining services by deception. We have already seen that the section 3 offence of the Theft Act 1978 has its own power of arrest, so all the offences discussed are ones for which hotel staff could arrest if they felt it to be necessary.

Index

Access to rooms, 80
Action in case of fire, 109–11
Aide-memoire, 51
Alarms, 130
Alarms, fire, 108–9
Alcohol:
  control, 68
  dispensers, 69
Aliens Order, 142
Arrest, power of, 146–7
ASIS, 9
Association of Hotel Chief
  Security Officers, 9

Banqueting, 74, 84
Bars, 68
Bar staff, policy, 69–70
Bellmen, 92–3
Bomb:
  attacks, 113
  threats, 6
    form for recording, 124–5
    threat calls, 114–16
Burglary, 6
Buying out, 69

Call girls, 95
Car parks, 89
  charges, 90
  security risks, 91
Car parking, 90
Cash:
  collection, 14
  handling, 25–9

point, 25
  procedures, 28
Cashier's office, 27
Casual staff, 74
Chambermaids, 49, 80
Chief Security Officer, 5, 7–11, 25,
  127–8
Cleaners, 84
Clock card systems, 38
Closed circuit TV, 18, 20, 22, 91
Connecting doors, 44
Contract security, 12, 14
Corridors, 22
Criminal Law Act 1967, 147
Customer fraud, 73

Damage to cars, 92
Deliveries:
  general, 40
  of food, 71
Disasters, 113–23
Display of wines, 75
Documentation, 66, 72–3, 76
Doors, 44
  frame, 44
  keeping shut, 81
  propped open, 81
Doormen, 93–4
Double locking, 48
Drunks, general, 32
Drunk:
  and incapable, 32
  and disorderly, 32

Electrical fence, 107
Electronic key system, 49
Emergency lighting, 105
Escape routes, 104
Evacuation, 116–18
Exterior entrance, 18–19

False alarms, 118
Fire:
  general, 97–111
  alarms, 108
  certificate, 101
  drill, 102
  doors, 103
  escape routes, 104
  fighting equipment, 106–7
  precautions, 6, 100–1
  staff training, 101–3
Firearm attacks, 121
Floor
  housekeeper, 83
  master key, 48
  safe, 28
  service, 66
Food and drink control, 65–78
Food and beverages, 70–1
Food deliveries, 71
Found property, 62–3
Fraud, 6

General manager, 10, 25
Guests:
  general, 31–41
  entry control, 37–9
  money, 25
  of hotel staff, 39
  property, 6
  rooms, 43–54
  valuables, 60
Grand master keys, 48
Guild of Hotel Chief Security

Officers, 9

Hall porters, 41
Health and safety, 100
Hoaxes, 120–1
Hotel:
  building security, 17–23
  design, 17–18
  property, 50
  Proprietors Act, 2, 37, 135–7
  switchboard, 94–5, 114–15
Housekeepers, 83
Housekeeping, 11, 79–87
House porters, 83

Identification of linen, 86
Immigration (Hotel Records)
  Order, 142
Industrial Professional Security
  Association, 9
Innkeepers Liability Act, 2
Institute of Industrial Security, 9
Institute of Professional
  Investigators, 9
Interior security, 20
Internal hotel training, 128–9
Investigation of losses, 51–4

Junior staff training, 133

Keys, 6, 44–5
Key:
  card, 46
  control, 46
  issue procedure, 46, 48
  lost, 46–7, 81
  master, 47–8
  safe deposit, 56–8
  spares, 47
Kidnapping, 121
Kitchen, issue to, 71

Laundry, 85
Letter bomb, 119
Liability:
  admission of, 53
  limiting legal, 55–6
Licensing Act, 35
Lien, right of, 143
Linen, 86–7
Linen room, 86–7
Locks, 44–5
  safe deposit, 59
Loss:
  guests' property, 139–41
  of linen, 87
Lost and found property, 60–3
Lost keys, 47
Lost property, 61
Luggage room, 6

Master keys, 47–8
Mixers, 69
Mortise lock, 45

Occupiers Liability Act, 141
Offices, Shops and Railway
    Premises Act, 1963, 101
Organising security departments,
    5–15
Outside contractors, 40
Outside exits, 105

Page boys, 92–3
Panic alarm buttons, 27
Parcel bomb, 119
Pay out points, 27
Pay packs, 27
Police hotel squad, 133
Police Review, 9
Property, security of, 55–63
Prostitutes, 35–6
Protection of escape routes, 104
Public:
  areas, 6
  entrances, 19

telephone boxes, 27

Race Relations Act, 138
Radio controlled bombs, 120
Restaurants, 72–4
Right of lien, 143
Right to refuse entry, 37
Room:
  cleaning, 80–3
  keys, 45
  security, 43–54
  service, 76-8

Safes, general, 26, 27, 28, 130
Safe deposit:
  bags, 59
  boxes, 56
  facilities, 56–60
  keys, 56–7
  log, 57
Safety, 6
Safety of guests, 141
Search, 116
Security:
  service, 76–8
  of car parks, 90
  of property, 55–63
  patrols, 11, 13, 22, 50
  policy, 6–7
  risks, 9
  training, 129–31
Security Gazette, 9
Security and Protection, 9
Security World, 9
Senior staff training, 131–2
Sex Discrimination Act, 138
Spare keys, 47
Staff:
  catering, 73–4
  entry control, 37–9
Submaster keys, 48
Stock reconciliation, 68
Suspicions, reporting of, 82
Switchboards, 94–5

Television sets, 95–6
Terrorism, 113–23
Theft, 6
  Act, 1968, 144–5
  Act, 1978, 145–6
  recording of, 54
Thieves, 33
Timekeepers, 38
Trade descriptions, 142–3
Training, 80–1, 101, 127–34
Traveller, 2
Trespass, 33, 145

Undesirables, 31–41

Valet service, 85
Valuables, deposit of, 60

Wages, 26
Walk-outs, 33–5, 48, 82
Window cleaners, 40
Windows, 21, 49–50
Wines, 74–5